Managing Complex Technical Projects

A Systems Engineering Approach

Second Edition

For a complete listing of titles in the
Artech House Technology Management and Professional Development Library,
turn to the back of this book.

Managing Complex Technical Projects

A Systems Engineering Approach

Second Edition

Ian Faulconbridge

Michael Ryan

ARTECH HOUSE

BOSTON | LONDON

artechhouse.com

Library of Congress Cataloging-in-Publication Data
A catalog record for this book is available from the U.S. Library of Congress.

British Library Cataloguing in Publication Data
A catalogue record for this book is available from the British Library.

Cover design by Iain Hill

ISBN 13: 978-1-68569-131-8

685 Canton Street
Norwood, MA 02062

10 9 8 7 6 5 4 3 2 1

Contents

Preface

The need to manage complexity is now commonplace in almost all fields of undertaking. Complex systems such as cars, airplanes, airports, financial systems, and communications networks commonly involve large numbers of disparate stakeholders, developers, operators, and customers. The development of major systems such as the Boeing 787 Dreamliner, for example, can take up to a decade involving millions of hours of work by a workforce of tens of thousands located in dozens of countries. At the same time, the need to accommodate rapid marketplace changes applies considerable pressure on traditional engineering disciplines to deliver suitable systems in a timely manner.

The principal role of project management is to bring these complex systems into existence while achieving quality, schedule, and budget goals. Although there have been significant efforts over the last two decades to define the discipline of project management and to improve the competency of project managers, projects still fail with alarming regularity. We have become accustomed to hearing of the difficulties associated with complex projects—cost and schedule overruns, dramatic failures to achieve requirements, and project cancelations. When examining the causes, it becomes apparent that such failure is rarely the result of the failure of one or more of the contributing disciplines; that is, it is rare that a project fails solely because business analysts do not understand the problem or opportunity to be pursued or that project managers cannot manage their schedule or their budget, contract managers can't manage contracts, engineers cannot design adequately, or production processes cannot produce a quality product.

In short, it is necessary that each of the associated disciplines pays due attention to their profession, but doing so is not a sufficient condition for project success. Education, training, and practice within each discipline are fundamental but will still not lead to project success without a major focus on the methodologies and processes through which those disciplines combine to develop, deliver, and support complex systems.

Complex technical projects can only be completed effectively by addressing the complete system life cycle. First, it is necessary to define business and stakeholder needs and requirements formally; then they need to be combined into an integrated set to provide a comprehensive description of the functionality (functional architecture) of the system to be developed. These functional needs and requirements are then analyzed and elaborated to create a functional description of subsystem requirements, which are then allocated to physical configuration items to provide a physical architecture of the system. The aim of developing the physical subsystems and configuration items is to decompose the complex system to a combination of well-defined elements that can be designed and then built by manageable teams using extant processes and procedures. The subsequent development of these separate physical elements must be managed, however, so that they are tested and integrated into the final system to be validated and delivered. To be successful, the entire process must be planned, recorded, and managed.

This book provides a basic but complete coverage of the management of complex technical projects and, in particular, of the discipline known as systems engineering through which that management is conducted. We offer a framework encapsulating the entire systems engineering discipline, clearly showing where the multitude of activities fits within the overall effort. The framework provides an ideal vehicle for explaining the many facets of the complex discipline of systems engineering.

We take a top-down approach that introduces the philosophical aspects of the discipline and provides a framework within which the reader can assimilate the associated activities. Without such a reference, the practitioner is left to ponder the plethora of terms, standards, and practices that have been developed independently and often lack cohesion, particularly in nomenclature and emphasis. The field of systems engineering is often viewed as dry, detailed, complicated, acronym-intensive, and uninteresting. Yet the discipline holds the key to delivering complex technical projects on time and within budget and to the quality expected, consequently avoiding many of the failures of the past. The intention of this book is both to cover all aspects of the systems engineering discipline and to provide a framework for the consideration of the many issues associated with engineering complex systems, particularly as they relate to the management of the projects that bring those systems into being.

Our secondary purpose is to describe a complex field in a simple, easily digested manner that is accessible to a wide spectrum of readers in a broad range of disciplines, from students to professionals, from novices to experienced practitioners. It is directed at a wide audience and aims to be a valuable reference for all professions associated with the management of complex technical projects: project managers, systems engineers, quality assurance representatives, integrated logistic support practitioners, maintainers, and so on.

In line with the top-down approach of systems engineering, we focus on the early stages of the system life cycle since the activities in these stages have the greatest impact on project management (i.e., on the successful acquisition and fielding of a system). In the interests of balance, however, we use the systems engineering framework to provide an overview of all other aspects related to systems engineering.

Chapter 1 is dedicated to the introduction of systems engineering and the development of the framework within which the remainder of the book is written. Chapters 2 to 4 examine in more detail the issues associated with the most important activities undertaken during early design—that is, during the first three phases of the system life cycle (conceptual design, preliminary design, and detailed design and development). Emphasis is given to the early activities associated with conceptual design and preliminary design. Chapter 5 deals with the broad topic of systems engineering management and details some of the activities normally associated with engineering management. Some of the more common and popular systems engineering tools are introduced in Chapter 6, which explains these tools' application to engineering management and processes. Chapter 7 explains the interrelationship between the systems engineering effort and other closely related disciplines such as business analysis, quality management, integrated logistics support management, and the specialist engineering disciplines. Chapter 7 also discusses the characteristics of the principal project approaches to system development: waterfall development, incremental development, evolutionary development, spiral development, and Agile methodologies.

Systems engineering is a broad discipline, and its application to different projects always requires individual and independent thought. There is never a single solution that will work for all projects, and there is rarely a solution that is either completely right or wrong. This book aims to introduce the main systems engineering issues to the reader to facilitate some of that individual and independent thought.

1

Project Management and Systems Engineering

1.1 Introduction

The principal role of project management is to meet project requirements (the project scope) to the desired quality within a given schedule and budget. In doing so, project management encompasses an extremely large body of knowledge that is documented comprehensively in many texts. Our interest here is not to repeat those details but rather to show how the management of a project is significantly improved by embracing a systems engineering approach.

We begin this introductory chapter with a summary of the project management knowledge areas that underpin the activities required to bring a system into existence. We then look at the nature of that system, examining the systems approach, the system life cycle, and the parties involved. We introduce systems engineering and its relevance and benefits, providing a framework that is used throughout the book. We conclude by noting the strong relationship between systems engineering and project management, primarily because every technical decision made as part of systems engineering activities has an impact on project management activities and vice versa.

Section 1.1 introduces project management, systems engineering, and the examples used throughout the book.

1.1.1 Introduction to Project Management

The Project Management Institute's (PMI) *Guide to the Project Management Body of Knowledge* [1] defines a project as:

> *... a temporary endeavor undertaken to create a unique product, service, or result.*

and project management as:

> *... the application of knowledge, skills, tools, and techniques to project activities to meet the project requirements.*

Although projects can yield a wide variety of unique products, services, or results, we are interested here in projects that bring engineered systems into being to solve a business problem or to take up a business opportunity. Consequently, we are interested in the relationship between project management and systems engineering.

A useful way to investigate that relationship is through the 10 project management knowledge areas defined in the PMI's *Guide to the Project Management Body of Knowledge,* which are summarized in the following paragraphs, including a brief note on the role of systems engineering in the activities defined within each knowledge area.

- *Project integration management* ensures that all elements of the project are properly planned and coordinated. The coordination involves making trade-offs between conflicting alternatives to meet the needs and requirements of the business and of the key project stakeholders. The key to integration management is the development of the project charter and then the project plan, which result from the project planning processes and form a governing approach to control and execution. The plan must be cognizant of the business constraints and organizational policies as well as the outputs from the project planning processes. Once the plan is in place, the project integration management function ensures that the plan is executed. Systems engineering contributes significantly to the development of the activities upon which the project plan is developed as well as to the subsequent integration effort.
- *Project scope management* is primarily concerned with ensuring that the project includes all the work required to complete the project successfully without including any additional or unnecessary work. The first stage in scope management is to define a formal scope as the basis of the subsequent effort. PMI's *Guide to the Project Management Body of*

Knowledge [2] references the use of systems engineering as fundamental to the completion of this task; almost all the activities we discuss in this book focus on developing and managing the scope of the system. Once the scope has been developed and approved, the project must be broken into smaller, more manageable components. This breakdown is commonly performed using a work breakdown structure (WBS). The scope and the WBS must then be formalized through an acceptance process to ensure that all parties are committed to the same scope and the same work. We show the important role that systems engineering has in defining the system requirements and then allocating them to the elements in the project WBS. Once approved, the scope management activity must take control of any changes to the approved project scope.

- *Project schedule management* is a key project management activity that is sometimes referred to as *time management.* To manage the project's schedule, the project manager must start with a comprehensive list of the work activities that make up the project. Tools such as the WBS assist in this identification, and, again, systems engineering makes an essential contribution. These activities are normally performed in a particular sequence that is dictated often by the systems engineering aspects of the project. For example, the activity "hardware/software integration" can only be performed following the activities "hardware test and acceptance" and "software test and acceptance." Other complications with respect to time management eventuate when activities have specific start and end dates that cannot be altered. Once the activities and their sequences have been determined, the durations for each of the activities need to be estimated. Based on these pieces of information, the project manager can determine the schedule for the entire project. It then becomes an exercise in schedule control as activities are added, deleted, or modified throughout the project. The project manager is assisted in the time management function by tools ranging from sophisticated, automated software packages to a simple reliance on expert judgment to predict activity duration. Systems engineering activities are the major source of information regarding WBS activities, in terms of their formation, progress, and completion.
- *Project cost management* ensures that the project is completed within the approved budget. Resource planning, the first task in establishing the budget, determines the resources required to complete the project activities. Resources include personnel, materials, and equipment. Based on the resources required for each activity, the project manager can estimate the costs associated with each activity and the total cost of the entire

project. Changes in the project scope will invariably impact the cost aspects of the project; these impacts need to be monitored and assessed on an ongoing basis. Tools are available to assist with cost management, but a great deal of the estimation and control effort relies on expert judgment and experience. Systems engineering provides critical input for the development of the WBS but also provides input to cost-benefit tradeoffs as well as investigations to determine the effect of any proposed change on project cost and schedule.

- *Project quality management* is essential in project management. The project manager must lead in the quality assurance effort to ensure that quality is taken seriously by all involved. The first stage in the process is to plan the quality management effort by identifying which quality standards are to be applied to the project and how these will be satisfied. The overall project performance must be evaluated regularly to provide users with confidence that the selected standards are being satisfied. The systems engineering efforts in establishing work packages, as well as in review and audit (particularly configuration audits), provide essential input to project quality management.
- *Project resource management* is not exclusive to project management but is, nonetheless, a vital part of the project management function, ensuring the effective use of resources across the project. Human-resource management guarantees the most effective use of project personnel. Often, one of the first tasks of the project manager is to acquire and organize the personnel required to perform the activities associated with the project. Once this has been done, the project manager must maintain the unity and direction of those involved with the project, which requires sound leadership and management skills such as negotiation, communication, motivation, and team building. In addition to human resources, the project manager is responsible for all other resources used or consumed to complete the project. Again, the extent, nature, and timing of the use of these resources is largely defined by systems engineering endeavors.
- *Project communications management* is a fundamental activity for project managers. Communication ensures that relevant project information is collected and disseminated to the appropriate personnel in a timely manner. The project manager must determine the information requirements of the business and of key stakeholders in the project and the preferred format for the information to be provided to them. The project manager then must ensure that the information is distributed as planned. A particularly important part of communications management is the report

generation associated with status reporting and phase or project closure. The role of communications management is worthy of emphasis here as a project management function to which systems engineering provides great value. The systems engineer is, in some respects, part management staff and part technical staff. Systems engineers, like business analysts, are therefore largely conversant with both technical and management languages and can act as very effective translators between the different layers within organizations. Systems engineers can also assist with horizontal communication between different design teams, to assist with technical issues such as design space consideration, interfacing issues, and technical conflicts. Of course, the major systems engineering reviews also provide a significant contribution to the communication of project information.

- *Project risk management* focuses on the risks associated with elements of project management, such as schedule and cost. Systems engineering supplies input to the risk management process in concert with the project manager, particularly regarding assessment of the technical risks associated with the project. Chapter 5 further discusses risk management.
- *Project procurement management* ensures timely supply of goods and services from outside the immediate project environment in support of the project. The initial step is to plan what products and services are required to complete the project and when they will be needed. These details should then be recorded. The project manager should seek quotes or offers from suppliers and select the best of the offers. In some cases, the procurement will be a one-time purchase, but often one or more standing offers or contracts will result. The project manager will be responsible for managing the resulting contracts. Once all aspects of the procurement are completed, the project manager is responsible for the closure of any extant contracts.
- *Project stakeholder management* is the process used to identify stakeholders and to develop, implement, and control an appropriate management plan. The PMI's *Guide to Project Management Body of Knowledge* defines a stakeholder as any individual, group or organization that is affected by, or can affect, the system of interest [3]. Since stakeholders can influence the system development, it is essential that they are identified and that appropriate mechanisms are established to engage with them throughout the project to ensure that their needs are met. From a systems engineering perspective, engagement with stakeholders is essential to the establishment of appropriate business and stakeholder needs and requirements as well as their transformation into system requirements.

1.1.2 Introduction to Systems Engineering

Systems engineering [4] methodologies and practices emerged from experience gained in the US Department of Defense acquisition programs of the 1950s. These programs often involved complex and challenging user requirements that tended to be incomplete and poorly defined. Additionally, most programs entailed high technical risk because they involved large numbers of different technical disciplines and the use of emerging technology. Following a number of program failures, the discipline of systems engineering emerged to help avoid, or at least mitigate, some of the technical risks associated with complex acquisition programs. Today, systems engineering provides the framework within which complex systems can be adequately defined, analyzed, specified, manufactured, operated, and supported. Systems engineering processes and methodologies have continued to develop since the 1950s and are widely applied to many of today's challenging acquisition projects. The modern systems engineering discipline is curated by the International Council on Systems Engineering (INCOSE), which defines the systems engineering body of knowledge as articulated in its *Systems Engineering Handbook* [5] against which it certifies systems engineers.

The focus of systems engineering is on the system as a whole, and on its whole life, with the maintenance of a strong interdisciplinary approach. Project management, business analysis, quality assurance, integrated logistics support, and the traditional design disciplines such as hardware and software engineering are but a few of the many disciplines that are part of a coordinated systems engineering effort.

Section 1.7 continues our introduction to systems engineering in more detail after we have considered systems, the systems approach, and the system life cycle; first, Section 1.1.3 quickly introduces the examples we use throughout the book to illustrate concepts.

1.1.3 Introduction to Examples Used Throughout This Book

Throughout the following chapters, we use examples wherever possible to illustrate and reinforce key concepts as we introduce them. To avoid duplication and assist with an understanding of the whole systems engineering process, we also use a couple of simple examples that flow through the whole book. In doing so, we do not intend to capture the entire design process. Rather, we have chosen examples that can be readily recognized by readers from a wide variety of disciplines and specialties, so that readers need not become domain experts in a particular field just to understand any particular illustration. Most readers can immediately understand the system context, the need, the functional, performance and verification requirements, the interface and integration issues, the technical performance measures, the functional-to-physical translation, the broad trade-off analyses, and the physical configuration items involved in the final design. It also should be noted, however, that we do not at all suggest that the aggregation of examples throughout the text represents an adequate or complete design for each respective example system; in fact, the inclusion of such detail would obscure the general lessons that are to be illustrated by the example. Additionally, please note that we often illustrate by leaving an aspect or an artifact of the system design in an unfinished state.

Example 1.1 Introduction to Aircraft Example

A large aircraft operator (ACME Air) has identified a need for a medium-sized aircraft to replace the aging platform that it currently operates over domestic routes and some short international routes.

Example 1.2 Introduction to Domestic Security Alarm Example

Another division of ACME Industries, ACME Alarms, has a business need to develop a domestic security alarm. The company proposes selling the alarm to the domestic market to compete in price and functionality for all forms of domestic dwellings such as houses and apartments. The alarm is to be capable of being installed by the customer, should they desire, and must be operable in a back-to-base monitored mode as well as a stand-alone mode.

1.2 What Is a System?

Before we address the project management and systems engineering issues associated with bringing a system into being, it is useful to consider what is meant by a system.

1.2.1 Definition of a System

A *system* is a complex set of many often diverse parts subject to a common plan or serving a common purpose; that is, a system is the whole (or the set) that results when a number of things have been grouped together in a particular manner, for a particular reason. ISO/IEC/IEEE 15288 [6] defines a *system* in its purest form as:

> *an arrangement of parts or elements that together exhibit behaviors or meaning that the individual constituents do not.*

We will see shortly that, from the perspective of the design of an *engineered system* (i.e., a system that we design deliberately and bring into being), it is more useful to use a slightly narrower definition that a system is:

a combination of interacting elements organized to achieve one or more stated purposes.

This latter definition implies that a system comprises internal *system elements* with *interconnections (interactions)* between elements, and the very act of explicitly identifying the system that we are interested in implies an external system *boundary* that separates the system from its environment.

As illustrated in Figure 1.1, when we draw the boundary around selected system elements, we define the *system of interest* (SOI), which consists of the *system elements* and their *interconnections* that exist within the defined *boundary*, as well as interconnections to *external elements* that exist in an external *operating environment* (or perhaps in a related system).

The purpose of the system, encapsulated in a single sentence, is called its *mission*. This sentence, which must be clearly stated by business management and stakeholders, represents the starting point of the design process and provides the basis for the ultimate test of the system's fitness for purpose once it has been fielded. In the broadest sense, the mission of the system is to provide a solution to a business problem or opportunity.

Note that, by being explicit in our definition, we have narrowed the term "system" in its general use to be an "engineered system." This is very important because it has two major implications, described as follows:

- When we refer to a system as being comprised of system elements that are *interconnected* to achieve the system's mission, we imply that all three of those principal aspects result from conscious choice. That is, we are

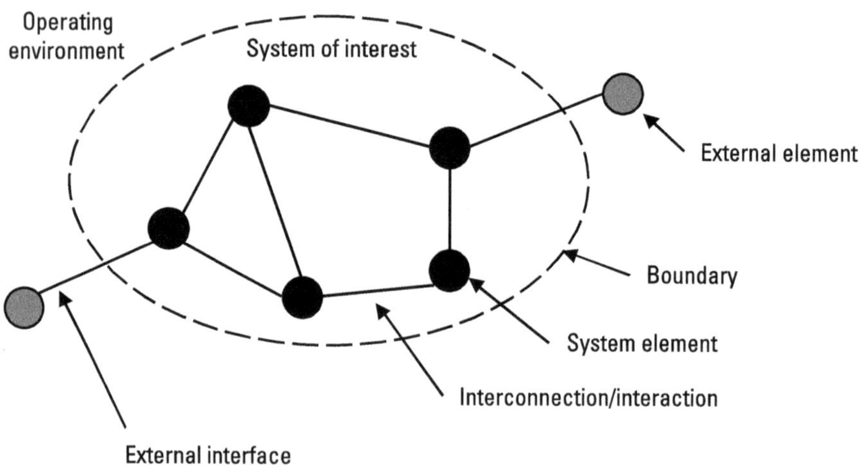

Figure 1.1 An SOI: its elements, interconnections, boundary, and interfaces to external elements in the system's operating environment.

referring to systems that have been deliberately *designed* (that is, *engineered*) and hence our interest in *systems engineering*.

- A system that has been engineered to perform a specified mission must be able to perform that mission with relative autonomy; that is, it must be managerially and operationally independent (and may well have been procured independently).

Although our consideration of a boundary has taken into account the system's relationship with external elements, sometimes we need to be cognizant of an even wider context so, as illustrated in Figure 1.2, an SOI might be considered as part of a wider SOI (WSOI) within an operating environment, which can be conceived as being part of a wider environment [7].

1.2.2 A System as a Capability—A Capability System

Recall that, in the broadest sense, a system provides a solution to a business problem or opportunity. A system is therefore much more than an aggregation of purely hardware or software products and combines a number of resources together in an organized manner. Consequently, a system must be described in terms of all of its constituent elements, including the major hardware and software products; the organization within which it will be fielded; the personnel who will interact with it in many ways; the collective training systems required, as well as the facilities, data, and support (including supplies); and the operating procedures and organizational policies required to keep the system in service. The system is fully defined by the combination of these resources operating

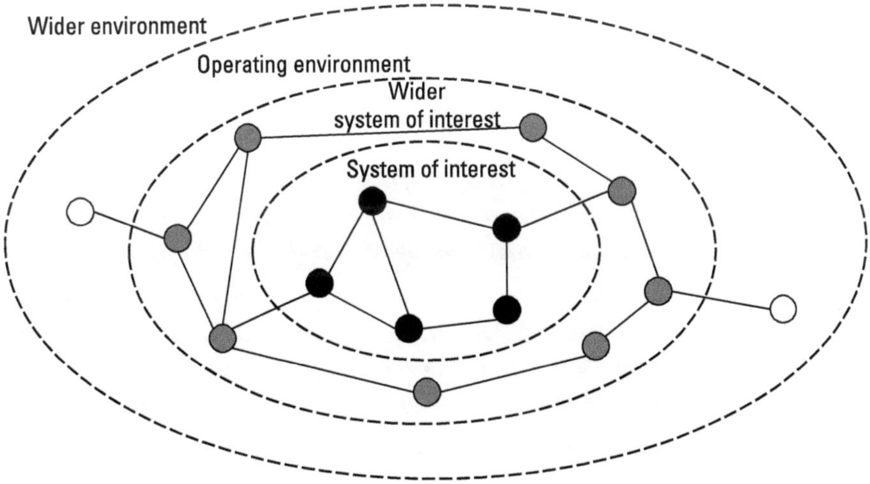

Figure 1.2 An SOI, which might be considered as being part of a wider operating environment.

in an operational environment to achieve some purpose. In that sense then, we could define a system as delivering an *operational capability.* It is common, therefore, particularly in defense environments, to refer to the system at this level as a *capability system* [8] and refer to its management as *capability management.* In business it is common to refer to the system at this level as the *asset* and to refer to *asset management* [9].

Having acknowledged all of the elements of a capability system (an asset), it must also be recognized that each of those elements will most probably have a different acquisition cycle—for example, people are "acquired" in a different manner to that in which the major equipment will be developed—and each element of the capability may even be acquired through a different acquisition element in the organization. The remainder of this text, for ease of description, focuses on the acquisition of the major equipment element of the capability system. We must always keep in mind, however, that this acquisition is being undertaken in parallel with the acquisitions of the other elements of the desired capability and that all the elements must be brought back together prior to introduction into service in order to field an operational capability/asset. Chapter 2 discusses this concept further and provides an illustration in Figure 2.5.

Example 1.3 Capability Elements for our Aircraft System Example

Capability elements for our aircraft system could include, but are not limited to, the following:

- Major equipment. *The most tangible part of the system is the hardware itself. The aircraft will be produced, distributed, and sold to operators who will then use the aircraft in a number of different ways, including for domestic and international operations. Software is also now a critical item within most systems. The aircraft is likely to use hardware and software to control a range of functions from engine management, through navigation and environmental control systems, to the communications and flight-control systems.*
- Personnel. *Aircrew need to operate the system. Groundcrew are required to maintain and support the fleet of aircraft. All these personnel will require individual training to some degree.*
- Support. *Maintenance facilities and equipment are required for routine maintenance and repairs throughout the aircraft's life. Materials required to operate the system include fuel, lubricants, and other consumables such as tires and spare parts.*
- Facilities. *Other facilities such as terminals are also necessary to operate the aircraft and its support systems.*

- Organization, policies, and procedures. *ACME must conform to a significant number of regulations and will need appropriate organizational structures, policies, and procedures to be able to operate the aircraft effectively.*
- Collective training. *Although air crew and ground crew will be trained for their individual roles, they will require collective training as an integrated team throughout the system life cycle.*
- Data. *Data is required to maintain and operate the aircraft. Data could include maintenance information such as specifications and drawings and operational information such as user manuals and instructions.*

1.2.3 Hierarchical Descriptions of a System

Recall that a *system* is a combination of *system elements* that interact to achieve a defined mission. Since each of these system elements will need to perform functions allocated to it so that it can contribute to the systems mission, we can consider the system to be a hierarchical composition of system elements [10], as illustrated in Figure 1.3.

The system elements in Figure 1.3 can be logical elements or physical elements, which support the concepts of a *logical architecture* and a *physical architecture,* as discussed briefly in Sections 1.2.3.1 and 1.2.3.2.

1.2.3.1 Logical Hierarchy

In a logical description of a system (see Figure 1.4), the system's mission is broken down into a hierarchical structure of its major functions. The logical description or architecture is therefore often called a *functional hierarchy,* or a *functional* (or *logical*) *architecture.*

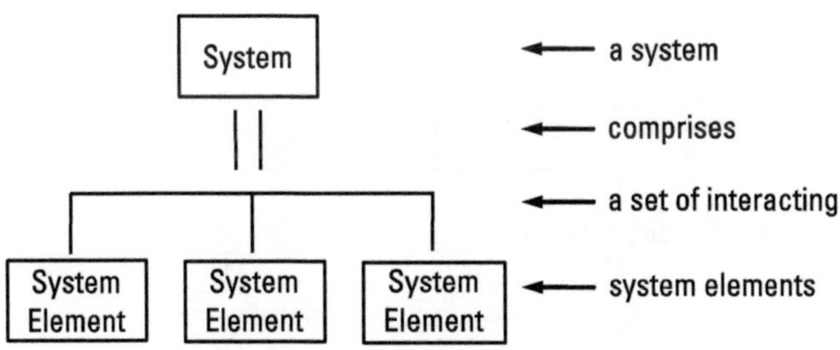

Figure 1.3 A system comprises a hierarchical set of interacting system elements.

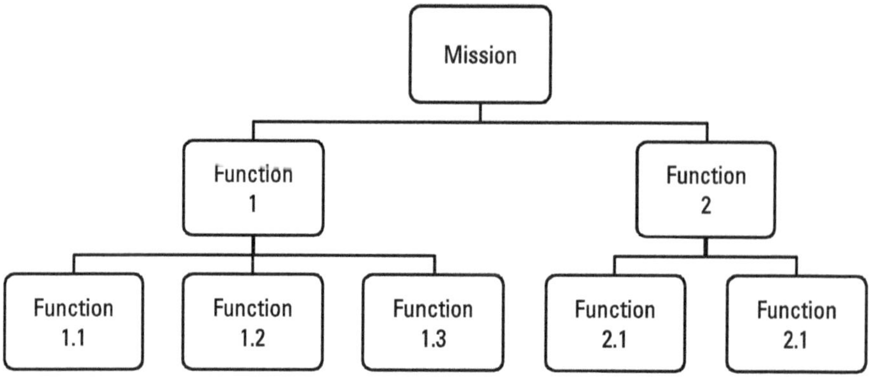

Figure 1.4 A logical (or functional) hierarchy of system functions.

1.2.3.2 Physical Hierarchy

In a physical sense, system elements are also normally described in a hierarchy. Here we use a four-layer hierarchy to illustrate the concept throughout the book. We describe a top-level entity known as the *system* that comprises a number of *subsystems* that comprise a number of *assemblies* that comprise a number of *components.* Although these terms are perhaps the most common, there are others in use. For example, Figure 1.5 illustrates what is perhaps the most complete physical hierarchy of system elements as defined by IEEE-STD-1220 [11], which also adds entities such as products, subassemblies, subcomponents, and parts to our simple four-layer taxonomy.

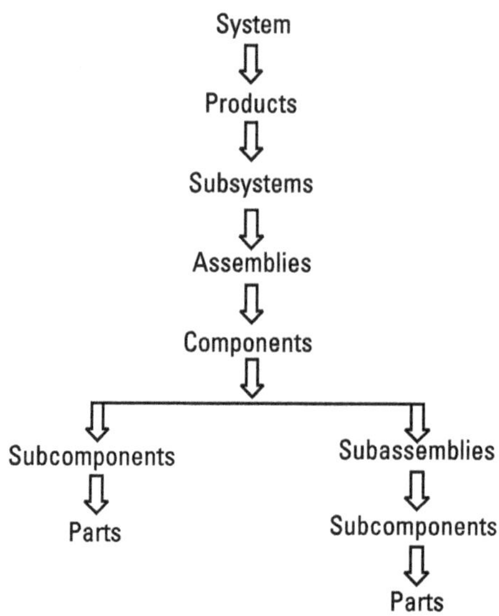

Figure 1.5 Hierarchy of physical elements of a system from IEEE-STD-1220.

It is worth noting here that in the context of systems engineering, the term "physical" is used to differentiate this hierarchy from the "functional" hierarchy described in Section 1.2.3.1 and that physical elements within a system may include elements of software that have been designed to operate on hardware elements to deliver prescribed functionality.

Another point of clarification is that application of these terms to specific situations and examples also depends significantly on the context of the situation and where within the overall project the system is being considered. For example, at the highest level of an aircraft, we would consider that the aircraft system contains, among other elements, the engine subsystem. The engine subsystem may consist of assemblies such as fuel tanks, pumps and lines, turbines, compressors, gear boxes, and hydraulic pumps. From the viewpoint of an engine manufacturer, however, the engine is commonly considered to be a system, with components including fuel, power plant, and hydraulic subsystems. The difficulty with considering the engine subsystem as a system in its own right is that an implicit part of the definition of a system is that it must be able to stand alone in its own right. By that definition, an engine is not able to be considered a system; it is only useful as an element of a system (that is, as a subsystem).

That is not a common view, however. For example, ISO/IEC/IEEE 15288 [12] considers an SOI to comprise a combination of interacting system elements, some of which may be systems in their own right, as illustrated in Figure 1.6. When the SOI consists only of system elements that are systems in their own right, the system-of-interest is commonly called a *system-of-systems* (SoS) [13].

Rather than extending the debate further here, we will simply continue with the notion that the system comprises subsystems that comprise assem-

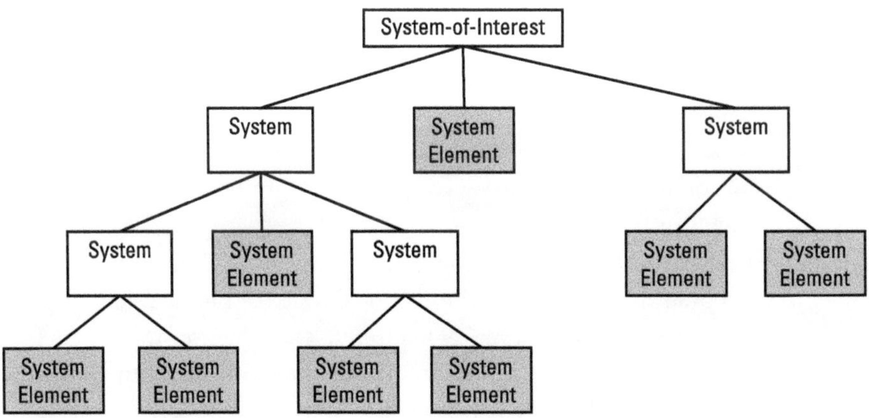

Figure 1.6 Hierarchy of elements of an SOI (after ISO/IEC/IEEE 15288).

blies that comprise components, and we will apply the systems approach (the systems paradigm) at each level. (See Section 1.3 for further discussion.)

1.2.4 Problem Domain and Solution Domain

Since a system can be considered as the solution to a problem or opportunity, it is common to consider the activities being undertaken throughout the life of the system to be in either the *problem domain* (*problem space*), where we use predominantly logical descriptions, or the *solution domain* (*solution space*), where we use predominantly physical descriptions.

Activities in the problem domain (including production of the logical architecture) are generally considered to be the responsibility of the customer/acquirer (the business owner); activities in the solution domain (including the physical architecture) are generally considered to be the responsibility of the organization implementing the system (the developer).

1.3 A Systems Approach

When considering the system, either from a logical or physical perspective, a *systems approach* focuses on the system as a whole, linking the value judgements of what is desired to the design decisions, which are based on what is feasible. Systems approaches draw on the wide range of knowledge appropriate to the SOI and its wider context to produce desired outcomes without adverse, unintended consequences. A systems approach, therefore, means that the design process remains focused on the problem/opportunity as well as the solution.

Systems approaches are often described as either *soft* or *hard,* defined as follows:

- *Soft systems approaches* are suited to addressing problems involving incomplete data and unclear goals drawing on subjective and humanist philosophies with constructivist and interpretivist foundations. Soft systems approaches comprise the following:
  - *Creative thinking* tools including brainstorming, morphological analysis, de Bono's six thinking hats, and Van Grundy's creative problem solving.
  - *Systems thinking* methodologies and techniques including cognitive mapping, influence diagramming, soft systems methodology, strategic choice approach, iterative planning, strategic assumptions surfacing and testing, the viable systems model (method), system dynamics modeling, rich pictures, systems thinking, causal loop diagramming, Actifeld, and field anomaly relaxation.

- *Hard systems approaches* are suited to solving well-defined problems with reliable data and clear goals, using analytical methods and quantitative techniques. Hard systems approaches comprise *systems engineering* and *engineering.*

Soft systems approaches and hard systems approaches are complementary. However, as Figure 1.7 implies, soft systems approaches are useful to assist in gaining and understanding of the problem space in early considerations; they focus on creative thinking, human perspectives, and systemic interrelationships. These methodologies focus on the understanding of problems in ill-structured, ambiguous, or socially complex situations. They involve stakeholder engagement, dialogue, and iterative problem structuring, so they are particularly useful in organizational change, policy development, and social systems. Techniques such as rich pictures, mind mapping, and causal loop diagrams help visualize multiple perspectives and emergent properties of systems, leading to adaptive and context-sensitive solutions.

In contrast, hard systems approaches rely more on structured, quantitative, and analytical techniques rooted in the systems engineering and engineering disciplines. These approaches focus on designing, optimizing, and managing well-defined systems with measurable parameters. They employ

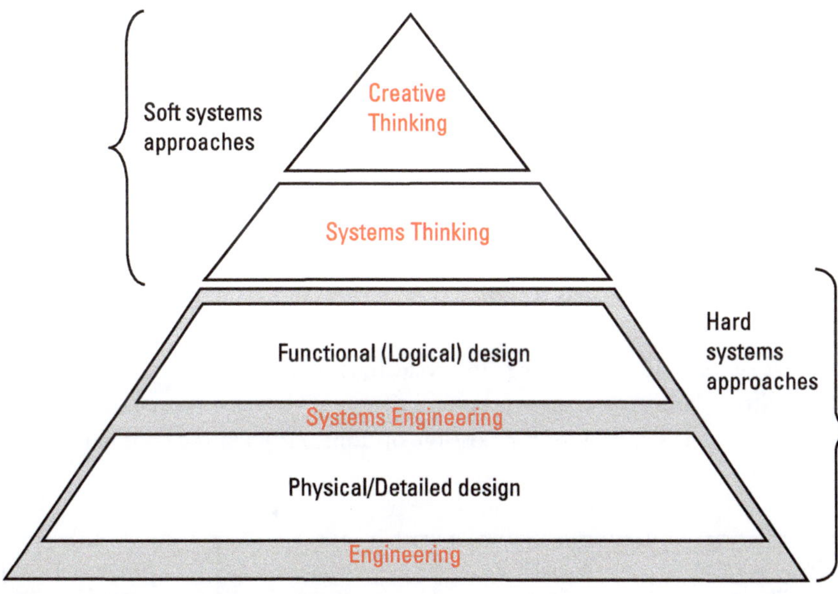

Figure 1.7 The hierarchy of soft and hard systems approaches: creative thinking, systems thinking, systems engineering, and engineering.

mathematical modeling and analysis; simulations; and rigorous methodologies, decision analysis, and modeling.

Together, these systems approaches enhance problem-solving capabilities. Soft systems thinking fosters creativity and adaptability, while hard systems thinking provides rigor and precision. Integrating both approaches enables a holistic understanding of complex challenges, balancing technical feasibility with social and organizational dynamics.

1.4 System Life Cycle

Now that we understand systems a little better, we need to consider that a system has a life. At some point in time the system doesn't exist; it is brought into being, it is operated and supported, and then it is disposed of once it can no longer serve the purpose (the mission) for which it was created. Throughout the life of a system there are, therefore, a number of phases and activities, each of which builds on the results of the preceding phase or activity. The sum of all these activities is called a *system life-cycle*, which can be described using a model that represents the conceptualization of the original business needs for the system and then its subsequent realization, utilization, evolution, and ultimate disposal [14].

There is no single accepted agreement on how many phases exist in a system life cycle or what those phases are called. In fact, there are a considerable number of generic life cycle stages in various life cycle viewpoints; for example, the INCOSE *Systems Engineering Handbook* [15] provides six example life cycles, each of which has utility in a particular organizational context. For the discussion in this book, we use the system phases illustrated in Figure 1.8, and for further detail during the Acquisition Phase, we use the life-cycle stages defined by Blanchard and Fabrycky [16] and MIL-STD-499B [17] and illustrated in Figure 1.9.

As shown in Figure 1.8, a generic system life cycle can be divided into four very broad phases, described as follows:

- *Preacquisition phase:* The life cycle begins in the preacquisition phase with an idea for a system being generated through business planning. Early consideration of the possible options results in the confirmation of

Preacquisition Phase	Acquisition Phase	Utilization Phase	Retirement Phase

Figure 1.8 Phases of a generic system life cycle.

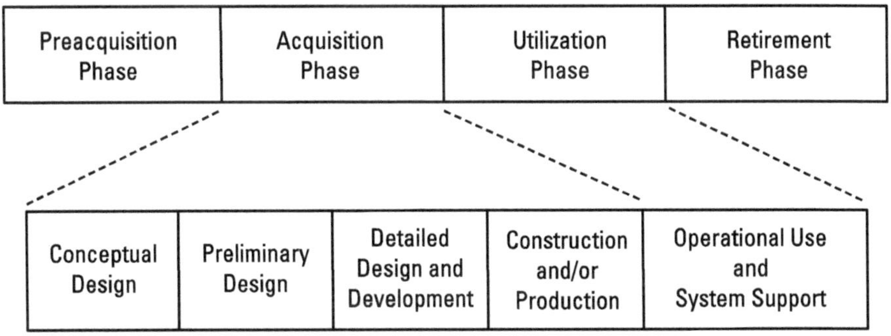

Figure 1.9 Activities in the acquisition phase and the utilization phase of the system life cycle (after Blanchard and Fabrycky).

the early business needs for the system, which are elaborated by a business case that justifies the expenditure of organizational resources on the acquisition and utilization of the system. In some instances, the preacquisition phase may determine that it is not feasible or cost-effective to proceed to acquisition (due to technology limitations or funding shortfalls, for example). In that sense then, the preacquisition phase is where organizations undertake a research and development effort to ensure that only feasible, cost-effective, justifiable projects are taken forward to acquisition.

- *Acquisition phase:* The business needs for the system provide the input for the acquisition phase, which is focused on bringing the system into being and into service. This would normally involve defining the system in terms of business needs and requirements, stakeholder needs and requirements, and system requirements and then engaging an implementation partner to develop and deliver the system.
- *Utilization phase:* The system remains in service during the utilization phase until the business has no further need for it, or it no longer can meet the functions required of it by the organization, or it is no longer cost-effective to keep it in service. During utilization, the system may undergo a number of modifications and upgrades to rectify performance shortfalls, to meet changing operational requirements or external environments, to enable ongoing support for the system to be maintained, or to enhance current performance or reliability.
- *Retirement phase:* Following operational use and system support, the system is eventually phased out and retired from service. The system life cycle concludes with the retirement phase. If the business need for the capability still exists in the organization, the conclusion of one system's

life cycle marks the start of another, and the process begins again. If the system has reached the end of its economic life, the life cycle ends in disposal.

It should be noted that the life cycle illustrated in Figure 1.8 shows the phases and activities in sequence and is not intended to represent any particular development model such as the waterfall, incremental, iterative, evolutionary, spiral, or Agile development models. These models (discussed in Section 1.9 and Chapter 7) represent different approaches to implementing the activities of the system life cycle in Figure 1.8.

Project management and systems engineering are predominantly related to the acquisition phase and, to a lesser extent, the utilization phase of the system life cycle. For these two major phases, we use the life cycle activities in Figure 1.9, which are based on those defined by Blanchard and Fabrycky [18].

1.5 Acquisition and Utilization Phases

As shown in Figure 1.9, the activities in the acquisition phase are conceptual design, preliminary design, detailed design and development, and construction and/or production. In the utilization phase, the activities are operational use and system support, which are undertaken in parallel. While there is no standard taxonomy, we choose these activities as a framework here because they are generally accepted in the systems engineering community for the following reasons:

- These activities emphasize that a system begins with perceived business needs and finishes with retirement and, ultimately, disposal—the so-called cradle-to-grave approach.
- There is a clear delineation between the acquisition and utilization phases of a system, which recognizes that there is a period of transition required as the system is transferred from the acquirers to the operators/users. Here that delineation is noteworthy for us because the project to acquire the system ends with the system's introduction into service—that is, at the end of the acquisition phase.
- The activities show the necessary detail in the early stages of the acquisition phase (particularly in conceptual design and preliminary design) where the application of project management and systems engineering methodologies and practices have the potential to make the most significant contribution.

- Importantly, within the acquisition phase, the activities also differentiate clearly between the problem domain, which contains the logical description of the system (the product of conceptual design), and the solution domain, which contains the physical description of the system (the products of preliminary design and detailed design and development) and then the realized system itself (the product of construction and/or production).
- Additionally, the separation of the early system design into conceptual design (what and why) and preliminary design (how) is very important since the responsibility in most programs transitions from the customer/acquirer for conceptual design to the contractor/developer for preliminary design.

1.5.1 Acquisition Phase

Each of the activities shown in Figure 1.9 for the acquisition phase is described in more detail in the following sections and chapters Figure 1.10 outlines the major tasks undertaken and the main artifacts produced in each activity of the acquisition phase.

1.5.1.1 Conceptual Design

Conceptual design focuses on the development of a set of clearly defined requirements, at the system level, and in logical (functional) terms. Although clearly defining the requirements of the system would seem a logical (and essential) first step, it is often poorly done and is commonly the direct cause of problems later in the development process. Business managers and stakeholders sometimes, understandably, prefer to describe their requirements in loose and ambiguous terms to protect themselves from subsequent changes in their needs and their business environment. The conceptual design process aims to

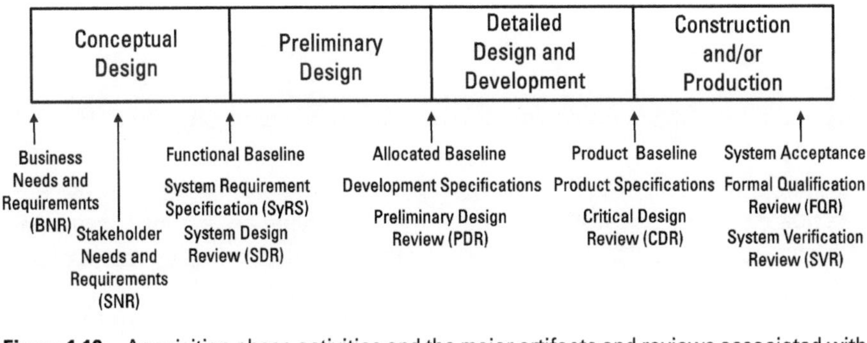

Figure 1.10 Acquisition phase activities and the major artifacts and reviews associated with each.

avoid this ambiguity by providing a formal process by which the business needs and requirements (BNR) [19] are articulated and confirmed by business management and then elaborated by stakeholders at the business operations level into a set of stakeholder needs and requirements (SNR) [20], which are further elaborated by business analysts and requirements engineers into a set of system requirements in the system requirement specification (SyRS) [21]. There may be one SyRS for the entire capability system, but it is more likely that there is one SyRS for each of the constituent elements of capability (i.e., major equipment, personnel, support, training, and facilities). Recall that each of these constituent capability elements may be developed independently, perhaps through separate contracts. Again, Chapter 2 discusses this aspect in more detail and illustrates the concept in Figure 2.5.

The BNR, SNR, and the SyRS are the key elements of what is called the functional baseline (FBL), which describes the whats and whys of the system. The FBL therefore contains a system-level logical architecture that meets the business and stakeholder needs and requirements.

Conceptual design ends with the system design review (SDR), which finalizes the initial FBL. The SDR provides a formalized check of the logical design; communicates that design to the major stakeholders; confirms external interface and interoperability issues; confirms the BNR, SNR, and the SyRS; and provides a formal record of design decisions and design acceptance.

1.5.1.2 Preliminary Design

The aim of preliminary design is to convert the FBL into an upper-level physical definition of the system configuration or architecture (the "hows" of the system). Preliminary design is therefore the stage where logical design is translated into physical design, where focus shifts from the problem domain into the solution domain. The result of preliminary design is a subsystem-level design known as the allocated baseline (ABL) in which the logical groupings defined in the FBL have been defined in more detail and then regrouped and allocated to subsystem-level physical groupings—elements of which are called configuration items (CIs)—that combine to form the upper-level physical design of the system. At the center of the ABL are a series of development specifications, which contain the subsystem-level requirements grouped by subsystem/CI as well as the interface requirements among those subsystems/CIs.

The ABL is titled as such because the system-level requirements that are logically grouped in the FBL are allocated at this next baseline into physical groupings (subsystems/CIs). The ABL therefore represents a subsystem-level architecture (couched in physical terms) that meets the requirements of the system-level architecture (couched in logical terms) contained in the FBL.

The initial ABL is formalized at the preliminary design review (PDR). The PDR ensures the adequacy of the preliminary design effort prior to focusing on detailed design. The PDR assesses the technical adequacy of the proposed solution in terms of technical risk and the likely satisfaction of the FBL. PDR also investigates the identification of subsystem/CI interfaces and the compatibility of each of the subsystems/CIs.

1.5.1.3 Detailed Design and Development

The ABL developed during preliminary design is used as the basis from which to complete the development of the individual subsystems, assemblies, and components in the system. Prototyping may occur, and the system design is verified by test and evaluation. The result of the detailed design and development process is the initial establishment of the product baseline (PBL), as the system is now defined by the numerous products (subsystems, assemblies, and components) making up the total system (as well as the requisite materials and processes for manufacturing and construction). The definition of the system at this stage should be sufficiently detailed to support the commencement of construction and/or production activities, which can be confirmed by a production readiness review.

The PBL is established at the critical design review (CDR), which is the final design review resulting in the official acceptance of the design and the subsequent commencement of construction and/or production activities. The CDR evaluates the detailed design; determines readiness for production/construction; and ensures design compatibility, including a detailed understanding of all external and internal interfaces.

1.5.1.4 Construction and/or Production

The final stage within the acquisition phase is construction and/or production. System components are produced in accordance with detailed design specifications in the PBL, and the system is ultimately constructed in its final form. Formal test and evaluation activities (acceptance tests) will be conducted to verify that the system is meeting its requirements as it is being developed [developmental test and evaluation (DT&E)] and that the final system meets the requirements in the SyRS [acceptance test and evaluation (AT&E)].

Construction and/or production, and the acquisition phase, ends with the formal qualification review (FQR) or system verification review (SVR), which provides the basis upon which the acquirer accepts the system from the supplier. The FQR/SVR is informed by the results of AT&E.

1.5.2 Utilization Phase

On acceptance from the supplier, the system moves into the utilization phase. Further testing in conjunction with all elements of the capability (such as equipment, people, processes, training, facilities, and support), known as operational test and evaluation (OT&E), validates that the system meets its original business needs and that the system is suitable to enter operational service. The major activities during this phase are operational use and system support as well as continuous OT&E activities to validate that the capability system still meets the business needs. Systems engineering activities may continue during the utilization phase to support any modification that may be required to rectify performance shortfalls, to meet changing operational requirements or external environments, to enable ongoing support for the system to be maintained, or to enhance current performance or reliability.

1.5.3 Retirement Phase

The system life cycle ends with retirement of the system in the Retirement Phase, which may well overlap with the introduction into service of the replacement system should the organization still require the capability. If the system is at the end of its useful life at the end of the life cycle, retirement will mean disposal, because the system cannot be reused in any way. Functions associated with retirement include transportation and handling, decomposition, and processing of the retiring system. A retirement concept should be developed during the early stages of the acquisition phase. If considered early, disposal and phase-out issues will form some of the criteria against which the system is designed (often called "design for disposability"). It is important, however, that system designers focus on retirement, as well as the more limiting issues of disposal; planning for disposal is important, but a system may be able to retire from several life cycles before it is ultimately disposed of at the end of life.

Example 1.4 The RAAF F-111 Aircraft

> *An excellent example of an extended life cycle is the F-111 aircraft operated by the Royal Australian Air Force (RAAF). Following a midlife upgrade in the 1990s involving both avionics and engine modifications, the RAAF had originally expected the F-111 to remain in service until 2020. However, the aircraft was retired earlier than planned as part of the transition to the F/A-18F Super Hornet, with its final withdrawal occurring in 2010. Given that the origins of the F-111 can be traced back to the 1950s, the life cycle of the RAAF F-111s still exceeded 50 years.*

1.6 Parties Involved

There are a number of parties involved throughout the system life cycle, and there is not any standard taxonomy for their names. This section introduces the nomenclature used throughout the remaining chapters.

The *customer* organization is managed by *enterprise management,* which sets the direction for the organization and for business units managed by *business management,* which is responsible for managing the activities conducted by the *operations* element of the organization (undertaken by the *operators*—sometimes called *users*).

The systems employed within the organization are acquired by the *acquisition* element (also called the *acquirer* [22] or *tasking activity* [23]) of the organization under the auspices of a *project* (managed by a *project manager*). Project managers are supported by a number of related disciplines including *business analysis, systems engineering, requirements engineering, specialist engineering disciplines, quality assurance,* and *integrated logistic support.* Operators are supported in their operation of the system by the *support* element of the organization, which supports, sustains, and maintains the system throughout its life. In addition to the operational, acquisition, and support staff, there are many others within the customer organization who have a stake in the successful implementation of the project. These *stakeholders* can include representatives from the management, financial, operations, supply, maintenance, and facilities areas of the organization.

The system is obtained from a *supplier* [24] (also called the *performing activity* [25]), which may deliver the system off-the-shelf or may develop it, in which case they are often called the *developer* [26]. The supplier may be internal to the customer (acquirer) organization. If the development of the system is undertaken in-house, the acquisition element of the organization (the acquirer or tasking activity) will engage with the development element (the developer or performing activity) to develop the system. It is increasingly common these days for the supply or development to be undertaken by an outside organization called a *contractor,* which is the entity responsible for supplying the system (perhaps by designing and developing it) to meet the customer requirements. The relationship between the customer and the contractor varies with each project, but, for each project, the relationship is formally defined by the terms and conditions of the *contract* between the two parties. In many cases the contractor is not able to perform all the work required and devolves packages of work to a number of *subcontractors.* The terms and conditions relating to this work are described in the relevant *subcontract.*

Responsibility for the various phases of the system life cycle is spread across the enterprise (or organization) within which the eventual system will operate. The initial preacquisition phase is the responsibility of enterprise management, which conducts business planning and establishes the business case for the projects required to support an organization. A project is then set up with a project charter providing authority to a project manager to expend organizational resources on the acquisition of the system. Systems engineering is an important discipline that is responsible to the project manager to perform the technical management of the project throughout acquisition and then to the capability (product or asset) manager during utilization. Once acquisition is complete, and the system is introduced into service, it is operated by the users and supported by the support element. Note that, rather than considering that each party has a role to play in a particular stage, it is best to consider that all parties are involved at all stages in the life cycle, with the roles and responsibilities of each party shifting in emphasis between stages.

1.7 What Is Systems Engineering?

The authoritative definition of systems engineering should come from INCOSE [27, 28]:

> *... a transdisciplinary and integrative approach to enable the successful realization, use and retirement of engineered systems, using systems principles and concepts, and scientific, technological and management methods.*

However, there is a wide range of systems engineering definitions [29–33], each of which tends to reflect the particular focus of its source as well as being a product of its time. Although each of these definitions has a slightly different focus, a number of common themes are evident, Sections 1.7.1 to 1.7.6 focus on these themes.

1.7.1 Top-Down Approach

Traditional engineering methods are based on bottom-up approaches in which known components are assembled into subsystems from which the system is constructed. The system is then evaluated for the desired properties, and the design is modified in an iterative manner until the system meets the desired criteria. This approach is valid and extremely useful for relatively straightforward problems that are well-defined. Unfortunately, complex problems cannot be solved effectively with the bottom-up approach.

Systems engineering takes a systems approach by addressing the complex system as a whole, which facilitates an initial understanding of the business and stakeholder needs and requirements as well as the subsequent analysis of the system and its interfaces. Once system-level requirements are understood, the system is then broken down into subsystems and the subsystems further broken down into components until a complete understanding is achieved of the system from top to bottom. This top-down approach is a very important element of managing the development of complex systems. By viewing the system as a whole initially and then progressively breaking the system into smaller elements, the interaction between the components can be understood more thoroughly, which assists in identifying and designing the necessary interfaces between components (internal interfaces) and between this and other systems (external interfaces). For example, Figure 1.11 illustrates the ANSI/EIA-632 approach to top-down development [34].

It must be recognized, however, that while design is conducted top-down, the system is implemented using a bottom-up approach, as illustrated in Figure 1.2. That is, the aim of systems engineering can be considered as twofold: (1) to provide a rigorous, reproducible process by which the complex system can be broken into a series of simple components and (2) to design those components using the traditional engineering bottom-up approach. Importantly, the other facet of systems engineering is, therefore, to provide a process by which components and subsystems can be integrated to achieve the desired system properties.

Note that, as discussed in Section 1.2.3, the terms *system, subsystem, assemblies,* and *components* are relative. Each system comprises subsystems that consist of assemblies that consist of components. Each subsystem, however, can be treated as a system in its own right, containing elements such as subsystems, assemblies, and components. Referred to by IEEE-STD-1220 as the *system paradigm,* this is synonymous with the systems approach introduced earlier. Consequently, while the building-block concept for top-down development is quite useful, it is often a source of confusion among novices if the associated terms are allowed to be used in a relative manner. Here, as outlined earlier, we avoid that confusion by maintaining distinctions between the terms system and subsystem while always keeping in mind the systems approach.

1.7.2 Requirements Engineering

The development of a complete and accurate set of system requirements is fundamental to project success and is a primary focus of the early systems engineering effort (recognizing, of course, that a complete description may not be possible a priori). The life cycle of a system begins with business needs and stakeholder needs, which are ultimately transformed into a large number of statements of requirement that form the basis for the logical design and that are subsequently elaborated further to form the physical architecture. These transitions must be managed by a rigorous process, called *business analysis* or *requirements engineering,* that is aimed at ensuring that all relevant requirements are included (and that all irrelevant requirements excluded). The establishment of the correct requirements is fundamental to the success of the subsequent design activities. Poor requirements cannot be rectified by good design, so it invariably follows that rigorous development of requirements is essential for the acquisition to be successful. Note that we are not referring here to any time period in particular; the same principle applies to a full development period in waterfall

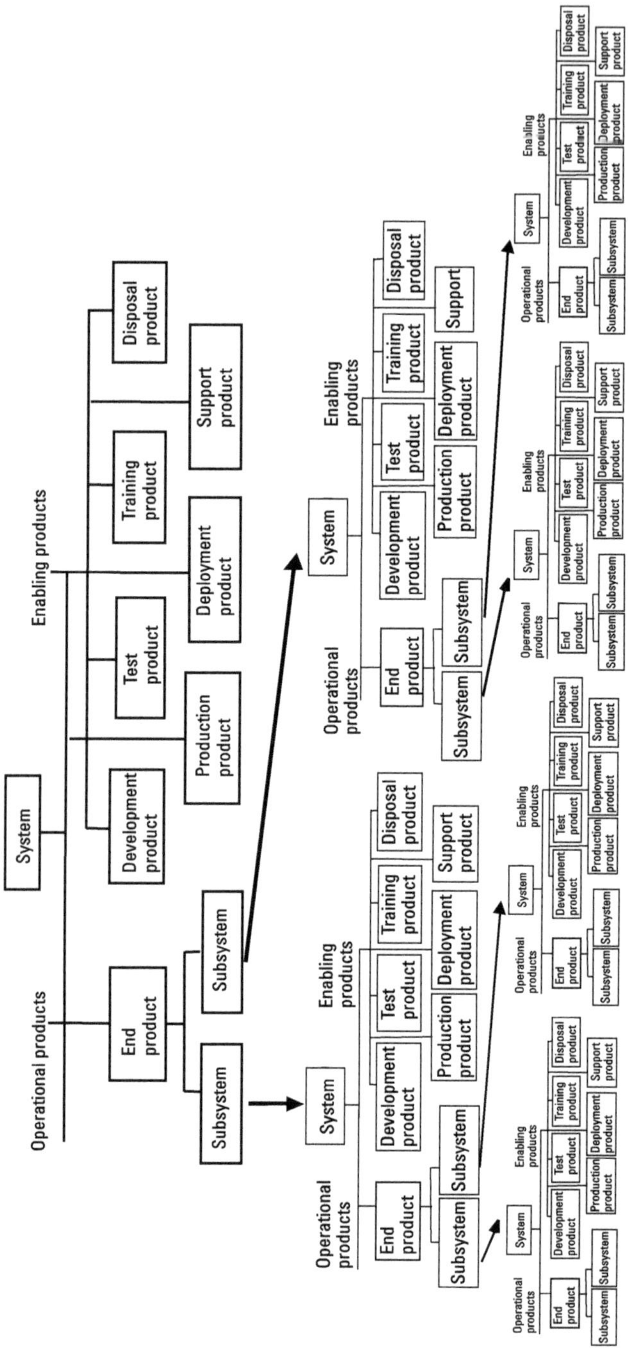

Figure 1.11 ANSI/EIA-632 building block concept for top-down development.

development, to a build in evolutionary development, or to a sprint in Agile. (See Section 1.9 and Chapter 7.)

Once a set of requirements has been established, the systems engineering process then focuses on the derivation and decomposition of these requirements from the system level down to the lowest constituent component (sometimes referred to as *requirements flowdown*). This process involves elicitation, analysis, definition, validation, and management of requirements.

Requirements traceability is also an essential element of the effective management of complex projects. Through traceability, design decisions can be traced from any given system-level requirement down to a detailed design decision (called *forward traceability*). Similarly, any individual design decision must be able to be justified by being associated with at least one higher level requirement (called *backward traceability*). This traceability is important since the customer must be assured that all requirements can be traced forward and accounted for in the design at any stage. Further, any aspect of the design that cannot be traced back to a higher level requirement is likely to represent unnecessary work for which the customer is most probably paying an unnecessary premium. Traceability also supports the configuration management (change control) process, especially the investigation of the impact of the change (see Chapter 5). Support for requirements traceability is a feature of the top-down approach, which provides a mechanism by which it can be guaranteed that requirements can be satisfied at any level. A bottom-up approach cannot provide the same guarantee.

1.7.3 Focus on Life Cycle

Systems engineering is focused on the entire system life cycle and takes this life cycle into consideration during decision making at any point. In the past it has been too common to consider solutions only in the light of the issues associated with the acquisition phase and to pay little attention to through-life support aspects. It is proper for project managers and their teams to focus on the acquisition phase of the project and on the development of a quality system that meets its scope while minimizing cost and schedule. However, a lack of attention to whole-of-life considerations can often lead to larger-than-expected costs in the utilization phase when budgets are insufficient to keep systems in service. A life cycle focus is on the capability system throughout its life cycle, not just on the product throughout its acquisition.

The significance of focusing on the system life cycle is that decisions made early in the acquisition phase are informed by activities proposed to be conducted later in the acquisition phase, and particularly in the utilization phase. For example, the design of an aircraft airframe must take into account the maintenance and operation of that airframe during the utilization phase. It would be

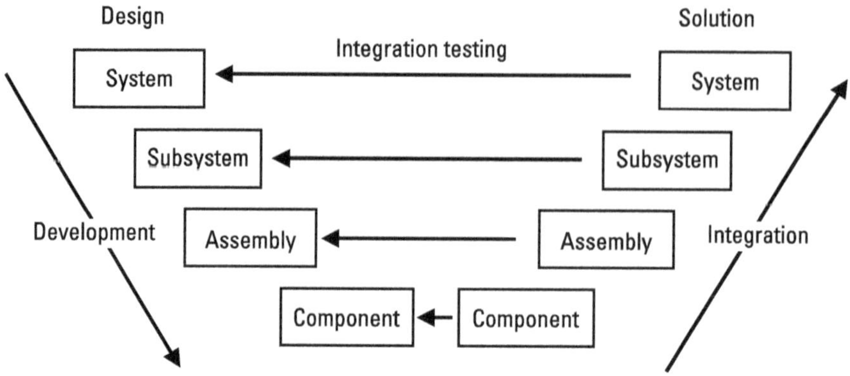

Figure 1.12 Bottom-up integration of a system: the system Vee diagram.

pointless to design the best airframe in the world if it did not have the necessary access points to allow maintenance personnel to service it or operators to operate it in the intended environment.

Other examples of utilization phase requirements that impact on equipment design or selection during the acquisition phase include reliability and availability. Reliability normally refers to the ability of the equipment to operate without failure for a given period. Availability is a measure of the degree to which a system is in an operable condition when required at some random point in time. Again, a superior design is pointless unless the system can meet specified minimum levels of reliability and availability and, for that matter, all the other "-ilities."

Economic factors provide arguably the most compelling evidence to support the focus on life cycle as opposed to product. In short, a life-cycle focus can save money in the long term. Experience has shown that a large proportion of total life-cycle cost (LCC) for a given system stems from decisions made early in the acquisition phase of the project. To that end, the maximum opportunity to reduce the total LCC, or total cost of ownership (TCO), of a system is presented in the acquisition phase. Poor requirements cannot be rectified by good design, so it invariably follows that rigorous development of requirements is essential for the acquisition to be successful. Put slightly differently, it is impossible for a project manager to manage scope, time, and quality to achieve project success if the scope of the project is fundamentally flawed; success or failure is therefore defined early in a project's life, knowingly or not.

Example 1.5 Life-Cycle Focus

As a simple analogy to demonstrate the concept (and problems) of a product focus as opposed to a life-cycle focus, consider a motor vehicle manufacturer that has introduced a new model. The new family sedan is a roomy, good-

looking, high-performance vehicle that costs half the price of similar models. Strong interest seems assured with buyers attracted by the vehicle's features as well as the cost-effectiveness of the purchase.

On further investigation, however, it transpires that the low purchase price has been achieved by a design that incorporates poor-quality components that require iterative, regular maintenance resulting in running costs that are five times higher than competing models.

This example continues to be relevant today where many people are choosing to switch from internal combustion engine vehicles (ICEVs) to battery electric vehicles (BEVs). At the time of writing, some of the older BEVs around the world have already reached the end of their original battery's useful life. The owners of these BEVs are discovering that the cost of replacing batteries in these BEVs is significant compared to the new car purchase price. This information is no doubt impacting on true life cycle cost of BEV ownership.

Clearly, a decision to purchase the new vehicle (whether it is an ICEV or a BEV) must focus on more than just the purchase price of the vehicle and must take into account the system life cycle aspects of operation, maintenance and support, and residual resale value.

1.7.4 System Optimization and Balance

A system architecture must represent a balance among the large number of requirements and constraints that cover a wide range of factors such as environmental, economic human factors, moral, ethical, social, cultural, and psychological. A simple, but essential, example is the balance between just two design factors: cost and performance; if either is allowed to dominate, it will generally be at the expense of the other.

A balance in system design must also be struck across the life cycle. Metrics such as cost-effectiveness must be measured across all phases, not just acquisition. It is often the case that savings made in acquiring the system are then countered by significant maintenance and repair costs in service. Although, as a corollary, it doesn't follow necessarily that paying more for a system during acquisition will guarantee that the support costs are lower.

Systems engineering performs a very important role in system design in ensuring that there is a balance among the various system components, because it is essential that optimization and balance are managed at the system level. As we discuss in Chapter 3, it does not necessarily follow that the combination of optimized subsystems leads to an optimized system. It is not normally useful, therefore, to allow the designers of subsystems to optimize their part of the system in isolation of system-level considerations; put another way, it is generally necessary for system-level considerations to constrain subsystem-level design in order for the system to be optimal.

It follows that a number of subsystems (if not all the subsystems) may need to be suboptimal (or at least constrained in some manner) to allow their combination (the system) to be optimal. A further advantage, therefore, of the top-down approach in systems engineering is that system optimization and balance can be achieved as a by-product of the design process.

Example 1.6 System and Subsystem Optimization

Consider the impact of incorporating an F1 engine into a small family saloon car; the engine may well be optimized for performance, but it will destroy the remainder of the drive train that has been designed for a much less powerful engine. Additionally, the F1 engine can propel the family car far faster than is safe given its suspension and brakes and much faster than is allowed by the legal speed limits. In general, if a subsystem (in this case the engine) is to be a useful servant of a system (in this case the car), its performance has to be defined (constrained) by the system designer, not by the subsystem designer.

1.7.5 Integration of Disciplines and Specialties

Systems engineering aims to manage and integrate the efforts of a multitude of technical disciplines and specialties to ensure that all needs and requirements are adequately addressed. Rarely is it possible for a complicated system to be designed by a single discipline.

Example 1.7 Integration of Disciplines

Consider our aircraft example. While aeronautical engineers may have a major role, the design, development, and production of a modern aircraft system requires a wide variety of engineering disciplines including electrical/electronics, safety and assurance, electromagnetic interference (EMI)/electromagnetic compatibility (EMC), production, metallurgical, and corrosion engineers. Other engineering disciplines are required for testing and for logistics and maintenance support as well as the design and building of facilities such as runways, hangars, refueling facilities, and embarkation and disembarkation facilities. Other nonengineering disciplines are involved in such aspects as marketing, finance, accounting, legal, and environmental.

In short, there could be hundreds, even thousands, of engineers and members of other disciplines involved in the delivery of a single aircraft system. Commonly, complex systems such as modern aircraft can involve millions of hours of work by thousands of people from a wide range of disciplines and backgrounds spread across dozens of countries.

> *The aim of systems engineering is to define the tasks that can be completed by these disparate disciplines and specialties and then to provide the management to integrate their efforts to produce a system that meets the business needs. In modern system development, this function is all the more important because of the complexity of large projects and their contracting mechanisms and the geographic dispersion of contractor and subcontractor personnel across the country and around the world.*

1.7.6 Management

While systems engineering clearly has a technical role and provides essential methodologies for systems development, it is not limited to technical issues and is not simply another engineering process to be adopted. Systems engineering has both a management and a technical role. Project management is responsible for ensuring that the system is delivered on time and within budget and meets the expectations of customers. The trade-offs and compromises implicit in those functions are informed by the products of systems engineering, which must be managed to be efficient. Additionally, the scope of the project is defined by the work breakdown structure, which is the result of the systems engineering effort. Systems engineering and project management are therefore inextricably linked. Section 1.14 discusses these issues in more detail.

1.8 Verification and Validation and Test and Evaluation

The entire systems engineering process aims to produce a system that is both verified against the descriptions produced during the systems engineering process and validated against the original needs, goals, and objectives that initiated the system development in the first case. Often these two associated aims are combined into the term *verification and validation,* although compounding the two into V&V is misleading and the practice is discouraged [35], since the two activities have quite different purposes. Verification ensures that we have "built the system right," and validation assures that we have also "built the right system" [36]. A well-managed approach to test and evaluation (T&E) aims to support the delivery of a system that is both verified and validated. Other systems engineering management functions—including technical reviews and configuration audits—also support the overall objectives of verification and validation.

There are three major categories of T&E that are applied to coincide broadly with verification and validation required in the activities of the acquisition phase, the transition between the acquisition and utilization phases, and the retirement phase. These categories are described as follows.

- *DT&E:* DT&E refers to the T&E activities undertaken during the acquisition phase to support the design and development effort. DT&E activities may also occur during the utilization phase to support activities such as the development of modifications. DT&E mostly supports verification activities to assure that the system element being developed is meeting, or has met, the design requirements of that system element.
- *AT&E:* AT&E represents the formal testing conducted to enable the acquirer to verify that the system meets the SyRS and to accept it from the developer. AT&E effectively forms the boundary or transition between the acquisition phase and the utilization phase and, as such, is an important contribution to FQR/SVR, which verifies that the delivered system meets the SyRS.
- *OT&E:* Once the system is accepted from the developer, OT&E may be conducted under realistic operational conditions by operational personnel in order to validate the capability system—that is, the major equipment combined with operational users, collective training, facilities, support, and other resources. OT&E is normally conducted for a period following acceptance of the system by the acquirer from the supplier (that is, after AT&E) and to validate that the system meets its BNR/SNR before introduction into service. OT&E is also conducted continuously throughout the system life cycle in order to determine continued fitness for purpose and to identify the need for any modifications and upgrades.

1.9 Systems Development Approaches

It should be noted that, although the generic system life cycle in Figure 1.9 shows the phases and activities in sequence, it is not intended to represent any particular development or acquisition model. Throughout the early chapters of this book, we describe systems engineering without discussing in great detail the development and acquisition context within which it might be undertaken. We present the life cycle phases sequentially, because it is the best way to explain the activities and artifacts of systems engineering. In doing so, we assume what is generally referred to as a *linear sequential* or *waterfall approach* to system development. There are, however, a number of other development approaches to implementing the activities of the system life cycle in Figure 1.9, such as the incremental, iterative, spiral, evolutionary, or Agile acquisition models [37], each of which has strengths and weaknesses depending on the nature of the system under development. The selection of a suitable development approach is a critical planning activity early in a system life cycle.

For simplicity in the early chapters, the linear sequential (waterfall) approach to system development is assumed in order to provide a logical, sequential flow of activities and deliverables that supports the provision of a readily understandable explanation of systems engineering. Additionally, the waterfall approach can be considered to be the basic building block upon which the alternative approaches such as incremental, iterative, evolutionary, spiral, and Agile development are built [38, 39]. A solid understanding of waterfall development is therefore useful, regardless of which other development methodology is employed.

Chapter 7, which considers systems engineering in the context of the major acquisition and development approaches, discusses these issues in more detail.

1.10 Systems Engineering Relevance

Systems engineering principles and processes are applicable (albeit to varying degrees) to a wide range of projects. For example, ANSI/EIA-632 [40] states that, as a systems engineering standard, it is intended to be applicable to the following:

> *...the engineering or the reengineering of:*
> *a) commercial or non-commercial systems, or part thereof;*
> *b) any system, small or large, simple or complex, software-intensive or not, precedented or unprecedented;*
> *c) systems containing products made up of hardware, software, firmware, personnel, facilities, data, materials, services, techniques, or processes (or combinations thereof);*
> *d) a new system or a legacy system, or portions thereof.*

Since it is difficult to imagine a system or attendant project that does not fit into the preceding description, it is always critical to understand the merits of systems engineering and apply them in a tailored manner, cognizant of the relative size, complexity, and risks associated with each undertaking. At one end of the spectrum are large complex projects making use of leading-edge developmental technology. These projects typically involve large sums of money, long time scales, and significant risks. At the other end of the spectrum are small projects making use of extant techniques and existing technology. These projects typically involve short periods, low costs, and a minimum of risk. Clearly different levels of systems engineering are applied to each of these types of projects.

The most obvious application of systems engineering principles and methodologies is in projects that are large and complicated. However, smaller and less complicated projects can also benefit from the application of systems engineering principles. In short, systems engineering practices, like project management practices, are always relevant but must always be tailored to be useful.

Example 1.8 Systems Engineering Relevance to Large Projects

A classic example cited by the Software Engineering Institute in its Systems Engineering Capability Maturity Model (SE-CMM) [41] demonstrates the need for good systems engineering. The Tacoma Narrows bridge was a very long suspension bridge with a flexible roadway that collapsed in 1940 due to strong winds that had set up an aerodynamic oscillation.

It is believed that the problems associated with the bridge could have been avoided or managed more effectively had systems engineering principles and processes been applied rigorously during the project. While adhering to good systems engineering practices would not necessarily ensure that designers are more likely to know the parameters of a particular problem, rigorous developmental approaches are more likely to draw out the relevant risk areas of a system (such as environment-related risks in this case).

The failures of system developments are often attributed to misunderstandings, ambiguities, misinterpretations, errors, and omissions in what the supplier is attempting to deliver. The result is a system that fails to address the acquirer's problem leading to a breakdown in the relationship between acquirer and supplier. Systems engineering specifically targets these problems and therefore is relevant to both parties in any system development.

Within an acquirer's project office, systems engineering is particularly relevant to the technical personnel who are responsible for the application of systems engineering principles as part of the overall project management effort. In most acquirer organizations, systems engineering personnel report to the project manager to apply the tools of the systems engineering to assist the project manager in each of the 10 project management knowledge areas. Systems engineering is therefore an essential element of the project manager's ability to acquire a quality system within budget, time, and scope constraints.

Suppliers should also be focused on developing processes and best practices to support the delivery of superior products and services to their customers. Systems engineering offers philosophies, methodologies, and tools that support the consistent development of quality products and services.

1.11 Systems Engineering Benefits

The principal causes of cost and schedule overrun on large-scale projects can be traced to such factors as overzealous advocacy, immature technology, lack of corporate roadmaps, requirements instability, ineffective acquisition strategy, unrealistic project baselines, project office personnel tenure and experience, and inadequate systems engineering [42]. This book focuses on the latter (which, incidentally, is the only one of those causes that is directly within the project manager's control during acquisition), because there are a number of potential benefits from the successful implementation of systems engineering processes and methodologies.

The first and most visible benefit is the scope for saving money during all phases of the system life cycle—LCC savings. While some may argue that the additional requirements imposed by systems engineering can increase costs, these increases are comparatively small and are generally felt in the early design phases. If applied appropriately, systems engineering can ensure that the savings achieved far outweigh the cost of implementing appropriate procedures and methodologies. Experience indicates that an early emphasis on systems engineering can result in significant cost savings later in the construction and/or production, operational use and system support, and disposal phases of the life cycle [43].

Systems engineering should also assist in reducing the overall schedule associated with bringing the system into service. Systems engineering ensures that the user requirements are accurately reflected in the design of the system helping to minimize costly and time-consuming changes to requirements later in the life cycle. If changes are required, they can be incorporated early in the design and in a controlled manner. The rigorous consideration and evaluation of feasible design alternatives during the design phases of the project promote greater design maturity achieved earlier.

System failures, cost overruns, and schedule problems are often the direct result of the poor application of business analysis and requirements engineering; poor requirements cannot be rectified by good design. The systems engineering discipline aims to put in place a rigorous process to produce well-defined needs and requirements, adequate levels of traceability between the different levels of technical design information back to the original needs and requirements, and system requirements that are both verifiable and consistent. This requirements-management process must achieve these results without presupposing a particular technical solution or placing unnecessary technical constraints on the solution.

Figure 1.13 provides a simplistic illustration of the impact of systems engineering on the system life cycle. Note that the greatest impact is through the rigorous application of processes and methodologies during the early stages of the project where the ease of change is greatest, and the cost of modification is the lowest. In fact, the curve in Figure 1.13 could be relabeled as the ease with which changes can be made throughout the system life cycle. Consequently,

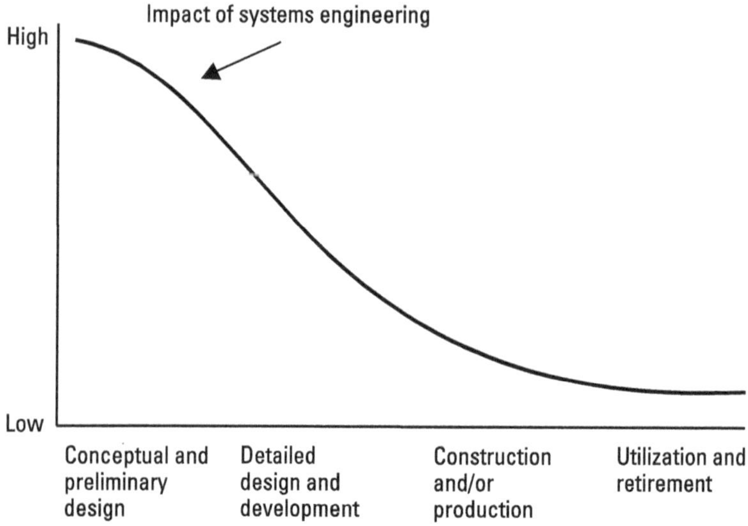

Figure 1.13 Impact of systems engineering on the system life cycle.

systems engineering provides the ideal opportunity to have the greatest impact on a project at a time when changes are easiest to make.

Systems engineering leads to a reduction in the technical risks associated with the system development. Risks are identified early and monitored throughout the process using a system of technical performance measures, and design reviews and audits. Design decisions can be traced back to the original needs and requirements, and any conflict can be identified and clarified early, significantly reducing the risk of failure later in the project.

Finally, and probably most importantly from a project management perspective, the disciplined approach of systems engineering leads to a delivered solution that meets the original intended purpose more completely; systems engineering makes an essential contribution to all the knowledge areas in the project management body of knowledge, which is detailed in Section 1.14.

1.12 Analysis, Synthesis, and Evaluation

All extant systems engineering standards and practices extol processes that are built around an iterative application of *analysis, synthesis,* and *evaluation.* The iterative nature of the application is critical to the systems engineering processes. Initially the process is applied at the systems level. It is then reapplied at the next level of detail and so on until the entire development process is complete. During the earlier stages the acquirer is heavily involved; in the latter stages, the supplier is mainly responsible for the continuing effort, which is monitored by the acquirer.

Prior to detailing the individual activities within the systems engineering processes, it is worth considering the basic foundations of the analysis-synthesis-evaluation loop illustrated in Figure 1.14. This concept is neither new nor terribly sophisticated; it is simply a good, sound approach to problem solving. While applicable in any domain, the loop is particularly fundamental to systems engineering. Sections 1.12.1 to 1.12.3 discuss, respectively, each component of the analysis-synthesis-evaluation loop.

1.12.1 Analysis

Analysis commences with a statement of perceived need or a set of customer requirements. During conceptual design, analysis investigates these needs and requirements and identifies the essential functions that the system must perform in order to meet the needs. Requirements analysis at the system level aims to answer the what, how well, and why questions relative to the system design. Analysis activities continue throughout the subsequent stages of the life cycle to help in defining lower level requirements, often called *decomposed* or *derived requirements* associated with physical aspects of system design.

A basic tool that can be used during analysis is the *functional flow block diagram* (FFBD), which shows the logical sequences and relationships of the functions of the system at the system level. The system-level functions detailed in the FFBD are further developed at subsystem level and below using more detailed FFBDs. Chapters 2 and 3 present examples of FFBDs. As the design becomes progressively more detailed, the system-level functions will be individually investigated and have an FFBD produced and so on until a very detailed set of functions and their interrelationships have been identified.

Once all the functions associated with the system have been identified, the requirements associated with each function can be defined. These requirements could include performance parameters such as speed, altitude, and accuracy; interoperability requirements detailing other systems with which the system under development must operate; and interface requirements to describe the necessary outputs expected from the system and the inputs to the system.

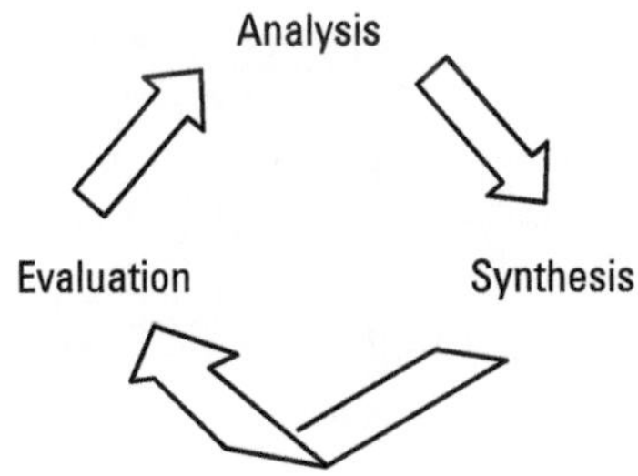

Figure 1.14 The analysis-synthesis-evaluation iteration.

Depending on the particular design phase, these requirements and functions may also be grouped in accordance with some sort of logical criteria and then allocated to a particular physical component of the system. That is, the component becomes responsible for the satisfaction of those requirements by performing the functions assigned to it. The allocation of requirements forms a description of the system elements and architecture and therefore assists in the process of synthesis or design.

1.12.2 Synthesis

Analysis revolves around what is required as well as how well and why. Synthesis, or design, then determines how. Synthesis is possibly the most widely recognized role of a professional engineer. Synthesis is the process where creativity and technology are combined to produce a design that best meets the stated system requirements. The term synthesis is more appropriate than design in the systems engineering context as it hints at the evolutionary nature of design and development.

In the early stages of the systems engineering processes, synthesis is limited to defining completely the functional design of the system and then considering all possible technical approaches using the results from the requirements analysis effort. From this consideration, the best approach is selected, and the process moves to the next level of detail. Later in the systems engineering processes, the selected design concept is synthesized further until, ultimately, the complete system design is finalized.

Although synthesis is the creative part of the systems engineering effort, a number of tools are available to the engineer to ensure that all alternatives are considered, and the most suitable alternative is ultimately selected. Chapter 6 discusses some potential aids to assist the designer during the synthesis process.

1.12.3 Evaluation

System cost and risk are directly associated with requirements and design. Evaluation is the process of investigating the trade-offs between requirements and design, considering the design alternatives, and making the necessary decisions. The process of evaluation continues throughout all stages of the systems engineering effort ultimately determining the system's satisfaction of original requirements. Trade-off analysis, discussed in Chapter 6, is one of the tools available to the system designer in performing evaluation of competing requirements or designs.

The outcome of the evaluation is selection or confirmation of the desired approach to design. Discrepancies are also identified if applicable and may result in further analysis and synthesis.

1.13 A Systems Engineering Framework

Discussions of systems engineering can quickly become complicated by the broad mandate of the system, the complexity and interrelationship of the many systems engineering activities, and the relationships with other disciplines throughout the system life cycle.

The ability to understand a complex subject such as systems engineering is greatly enhanced by a solid framework within which concepts can be considered. Examples of such a structure include the *Project Management Body of Knowledge* [44], which provides a clear framework within which to consider the many facets of project management, and the *Software Engineering Body of Knowledge* [45], which performs a similar role for the software engineering discipline. Without an equivalent framework, the broad scope of systems engineering soon becomes confusing given the complexity of its components and their many interrelationships. There are a number of excellent systems engineering standards available today that contribute to the elements of a suitable framework, but each standard contains complexity, (differing) terminology, and detail that requires substantial interpretation. The entry level of many students, junior engineers, and project managers therefore does not readily allow the use of such standards as effective frameworks within which to examine systems engineering.

In response to this conundrum and in an effort to provide a structure within which the systems engineering discipline can be more clearly understood and implemented, we have synthesized a systems engineering framework [46] through a thorough survey of existing systems engineering publications and standards and our experience in teaching systems engineering at a range of levels. Figure 1.15 illustrates this framework.

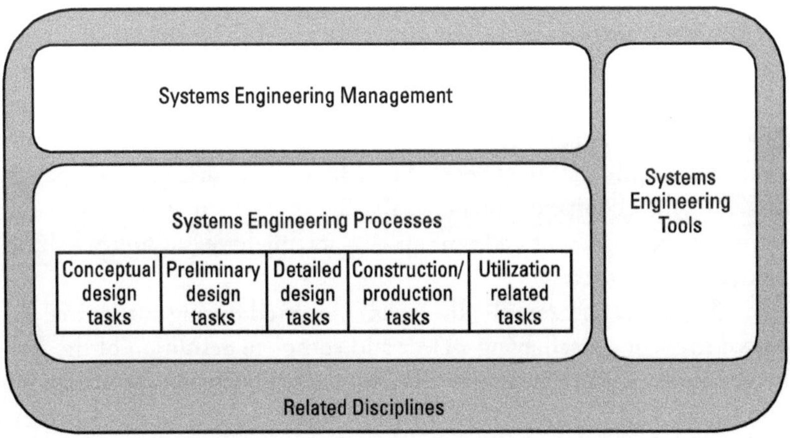

Figure 1.15 A framework for the consideration of systems engineering.

The framework illustrates the relationship of the three main elements of *systems engineering processes, systems engineering management,* and *systems engineering tools* and places them in context with *related disciplines* such as traditional engineering disciplines, project management, integrated logistics support, and quality assurance. The terms systems engineering management and systems engineering processes are sometimes used interchangeably. Here we make a distinction between the two.

We present the systems engineering processes as the hows of systems engineering, the application of which forms the foundation of the systems engineering effort. Over the top of these processes sits the systems engineering management functions, which are responsible for directing the systems engineering effort, monitoring and reporting that effort to the appropriate areas, and reviewing and auditing the effort at critical stages in the entire process. These two elements are supported by a range of tools, and all elements exist in the context of related disciplines.

The systems engineering framework provides an excellent structure within which to examine and explain the complex discipline of systems engineering. Experience in undergraduate and graduate courses as well as professional education short courses [47] has shown that the framework provides an excellent means of communicating the complexities and interrelationships of the systems engineering discipline, particularly to those who do not have a significant amount of project or systems experience. Those new to the discipline can successfully grasp the fundamental concepts of systems engineering within a relatively short timeframe.

1.13.1 Systems Engineering Processes

The basic systems engineering process is the analysis-synthesis-evaluation loop described in Figure 1.14. This loop is applied iteratively throughout the system life-cycle. Other systems engineering processes and tasks are divided into the life cycle stages within which they typically occur. This book does not attempt to detail exhaustively all systems engineering processes. Instead, we concentrate on the intent and main aim of each phase of the life cycle and examine some of the likely techniques that may be used to achieve that aim. We place particular emphasis on the acquisition phase of the life cycle, as it is the phase during which project management and systems engineering have the ability to have the most impact on a system.

In Chapter 2, for example, the tasks completed during conceptual design are shown to focus on achieving a clear and complete definition of the system-level requirements. We investigate the processes of articulation of the system mission, goals, and objectives and the subsequent integrated set of needs; system feasibility analysis; system requirements analysis; and system synthesis.

Chapter 3 investigates the processes associated with preliminary design, and Chapter 4 describes those related to detailed design and development and construction and/or production.

1.13.2 Systems Engineering Management

Systems engineering management is an overarching activity responsible for directing the systems engineering effort, monitoring and reporting that effort to the appropriate areas, and reviewing and auditing the effort at critical stages in the entire process; such management is the key to success in entire systems engineering and project management efforts. Chapter 5 addresses the major systems engineering management elements of technical reviews and audits, test and evaluation, technical risk management, configuration management, the use of specifications and standards, integration management, and systems engineering management planning. The preeminent position of systems engineering management in the framework illustrates that it is key to the entire systems engineering effort.

1.13.3 Systems Engineering Tools

Many tools exist to assist systems engineering processes and management, including a range of techniques and methods and a number of standards. Chapter 6 describes the most popular tools and standards under the categories of management tools and process tools. It is not the intended purpose of this book, however, to repeat information contained elsewhere in the many standards and artifacts; we simply demonstrate these tools' and standards' utility in systems engineering and project management.

Throughout the book, we present generic process tools such as requirements breakdown structures (RBSs), FFBDs, WBSs, trade-off analysis, prototyping, simulation, and model-based systems engineering (MBSE), as examples of tools that may be applied to the systems engineering process effort. Chapter 6 also describes the use of standards as systems engineering management tools, reviewing and summarizing MIL-STD-499B [48], EIA/IS 632 [49], IEEE-STD-1220 [50], EIA 632 [51], and ISO/IEC/IEEE 15288 [52].

1.13.4 Related Disciplines

There are many disciplines (both technical and nontechnical) related to systems engineering. In addition to project management, major examples include business analysis, logistics management, quality assurance, requirements management, software engineering, hardware engineering, and interface engineering (or integration engineering).

The relationship between the related disciplines and the various facets of systems engineering within a project depends very much upon the discipline in question. Some (such as project management) oversee the whole systems engineering discipline during acquisition, while others (such as hardware and software engineering) sit between systems engineering management and the processes, and others (such as quality assurance) sit alongside the systems engineering effort. Chapter 7 discusses these disciplines and their relationship with systems engineering.

1.14 Systems Engineering and Project Management

The strong relationship between project managsement and systems engineering is demonstrated by considering the 10 project management knowledge areas summarized in Section 1.1.1. Kauffman [53] investigated this relationship and demonstrated a significant intersection between the two professions by referring to PMI's for project management, and IEEE-STD-1220 and EIA/IS-632 for systems engineering when those standards first appeared in draft form. Figure 1.16 provides a summary of Kauffman's findings, augmented for project

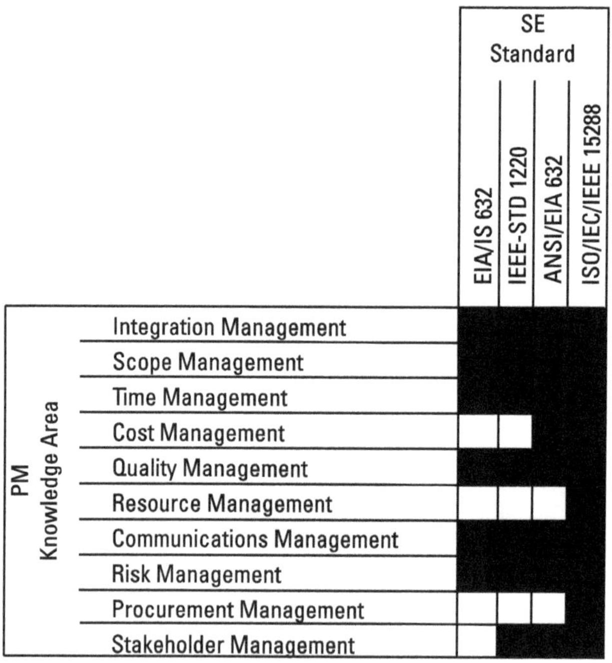

Figure 1.16 Relationship between the 10 knowledge areas of project management and systems engineering as defined in the major systems engineering standards.

management to include all current knowledge areas of the *Project Management Body of Knowledge* and for systems engineering to include the current versions of both EIA-632 and IEEE-STD-1220 as well as ISO/IEC/IEEE 15288.

In the earlier standards—EIA/IS-632, IEEE-STD-1220, and EIA-632—there are mostly areas that overlap between the two disciplines, but there are notable exceptions in activities traditionally considered the sole province of project management: cost management, resource management, and procurement management. Yet, there are still strong relationships even in those aspects. For example, project management is responsible for management of cost and schedule, although this can only be meaningful with input from systems engineering in terms of the development of the WBS.

However, Figure 1.16 illustrates that, over the last two decades, there has been a considerable change in thinking associated with the discipline of systems engineering. Note that in ISO/IEC/IEEE 15288 (the current premier systems engineering standard), the interrelationship between all aspects of both disciplines has been strengthened even further, because the standard not only refers to every aspect of the *Project Management Body of Knowledge* but also covers all life cycle aspects of systems development that are outside the acquisition phase and therefore outside the discipline of project management.

This also reflects major changes in relevant standards and the systems approaches to the system life cycle. As illustrated in Figure 1.17, the older systems engineering standards such as EIA/IS-632 and IEEE-STD-1220 viewed systems engineering as that subset of project management activities that focused on design and development of the system in support of acquisition; in quite considerable contrast, the current ISO/IEC/IEEE 15288 shows that project management relates to that subset of the systems life cycle processes that focused on the acquisition phase. Systems engineering focuses on the remainder

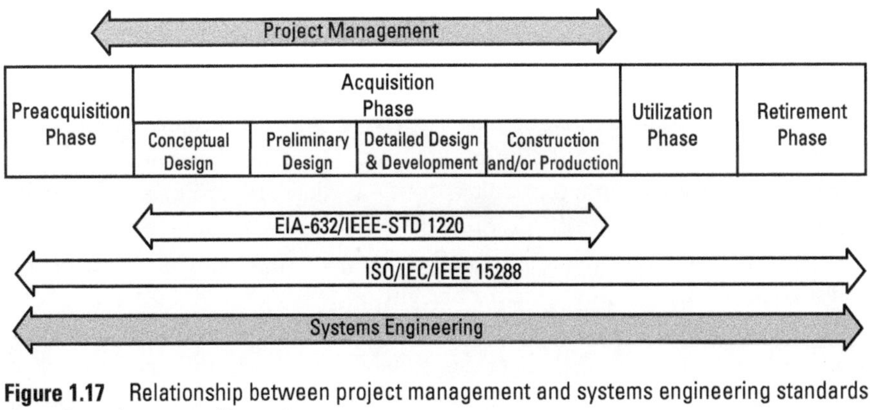

Figure 1.17 Relationship between project management and systems engineering standards throughout the system life cycle.

of the life cycle processes as well as the acquisition phase—that is, the entire life-cycle.

Whichever way they are viewed, however, Figures 1.16 and 1.17 illustrate that project management and systems engineering are inextricably entwined, particularly in the acquisition phase.

1.15 Summary

This chapter introduces project management and systems engineering, the systems approach, the nature of systems, the logical and physical descriptions of a system in the problem domain and the solution domain, the system life cycle and its constituent activities, and the parties involved in system acquisition. We also examine some definitions of systems engineering and extract the common themes of top-down design, requirements engineering, focus on life cycle, system optimization and balance, integration of related disciplines and specialties, and management. We then consider the benefits of systems engineering and examine the analysis-synthesis-evaluation loop, which is applicable from the basic engineering activities to the systems engineering activities right across the system life cycle. We conclude by noting that systems engineering and project management are inextricably linked.

Next in Chapter 2, we begin to detail each of the major systems engineering processes as part of a project as part of the acquisition phase. We begin with the first stage of the acquisition phase—conceptual design.

1.16 Some Qualifications Before We Continue

Before we continue, however, we need to be clear about the scope of our coverage throughout the book.

A project, by definition, produces a unique output. So, while there are general principles involved in the conduct of projects, the constituent activities of an individual project must be tailored into a relatively unique combination for that project to produce the unique output. There are therefore arguably as many ways to implement a particular project as there are unique systems to be developed. We cannot possibly address here the many combinations of factors associated with complex technical projects, and it would not be useful to attempt to do so because any benefit of an introductory overview would be lost. Although we are acutely aware of these complications, we just do not have the room to address them directly here. So, to navigate the multidimensionality of the project space in a useful manner, we take a direct path and make a number of simplifications which should be kept in mind when reading the following chapters; these simplifications are described as follows.

- Most organizations involved in systems development will incorporate a number of development methodologies (perhaps at least a dozen) as part of their systems engineering and project management policies and procedures. Then, as part of the early project management activities, organizational practice requires a project manager to justify the selection of an appropriate methodology as part of their development of a project management plan. Here we use the linear sequential development methodology (commonly known as waterfall development) as a framework as we progress through the chapters. Subsequently, Chapter 7 describes other major development methodologies (incremental, iterative, evolutionary, spiral, and Agile). There are two main reasons for taking this approach in an introductory text, outlined in the following:
  - First, the various elements of systems engineering are much easier to understand on first contact if they are explained sequentially; the chapters are therefore structured that way. However, just because we have explained the processes sequentially does not mean at all that we suggest that steps must be implemented that way. For each unique project output, an appropriate project development methodology must be chosen, tailored, and applied.
  - Second, even when a more elegant development methodology is applied, an understanding of the logical steps is essential; that is, it doesn't matter whether the project team is currently in a phase in waterfall, an increment in incremental, a build in evolutionary, a spiral in spiral, or a sprint in Agile. The steps we take the reader through here still have to be applied to the fundamental time period of the approach (phase, increment, build, spiral, or sprint).
- The word complex is included in the book title, and we refer to complexity throughout the chapters. As we move through the processes, we are focused on the methodologies, tools, and techniques that can be applied to project activities to deal with complexity in the project output. We are not, however, focused on complexity per se and do not address it directly.
- In any system development, there will be a number of what most modern standards call acquirer-supplier arrangements. These are also called acquirer-developer, customer-contractor, and tasking-performing activity arrangements, among a number of other terms. It does not matter whether the development is conducted, as are most government acquisitions, through a major formal tendering/contracting process or, as many in-house developments are, undertaken between two or more groups internal to the organization, or in any other arrangement. In all cases

(albeit slightly less formally for in-house development), it is essential that the acquirer-supplier arrangement is clear and unambiguous and that there are agreements for milestones, deliverables, budget, schedule, review and audit, verification, and validation, among other factors. This remains true (albeit less formally) even for short development periods such as sprints in Agile development methodologies. Consequently, since there can be as many unique contracting arrangements as there are projects bringing unique systems into being, we focus our explanations on the most formal of those arrangements that may be implemented between a large government acquirer and a major supplier. Once the reader is familiar with the issues that drive formal developments, it is relatively straightforward to apply the general principles to less formal arrangements, such as those guiding in-house developments.

- One of the most dynamic areas in modern systems engineering is in digital engineering (DE) and the associated MBSE. Here we focus on project artifacts that could be documents, spreadsheets, or databases and on tools that could be word processors, requirements management tools (RMTs), or models in MBSE. For example, we say that system requirements are recorded in the SyRS, a term containing the word specification, which might be construed to imply document. However, we do not use the term in that way. Rather, we expect that the SyRS will be recorded in any form suitable to the organization: from a document to a spreadsheet, a database, a model, or any other digital form, or some combination. Accordingly, when we use the term document as a verb, we mean to record as either a paper-based document, a database, or a model. Chapter 6 briefly describes MBSE, but we avoid direct reference to DE and MBSE in the remainder of the text for the following two reasons:
  - First, many projects still use paper-based documents to record project activities and will do so for years, or even decades to come; in fact, the authors have worked with a number of major international companies that still use word processors to capture requirements, contracts, and design outputs. While DE should be, and is being, embraced enthusiastically wherever possible, there is significant associated overhead, which requires significant organizational investment. Additionally, with major government contracts that can span a decade or more, many projects will remain in the nondigital or part-digital world for some considerable time.
  - Second, although the form of the physical artifact changes between the paper-based and the digital environments, the logic of activities

and relationships does not change; the same principles apply in both environments. Here, therefore, we focus on these principles and logical steps. For convenience, we describe the resultant artifacts as a logical result of each stage, phase, and activity. We do not at all propose that those artifacts take any particular physical or digital form; that is a matter for the project and the organization it supports.

- Huge strides have been made in the 2023 revisions of ISO/IEC/IEEE 15288, the INCOSE *Systems Engineering Handbook,* and its attendant manuals and guides to provide a common view of systems engineering in terms of life cycle processes, methodologies, tools, techniques, and terms. However, there are still many subtle (and sometimes not-so-subtle) differences in the ways in which systems engineering is applied in organizations around the world. We are therefore careful to provide a balance that reflects the older terms and processes as well as the new; our interest is for the content of the book to be as relevant to system developments that are just beginning as it is to those that have been in progress for a decade.
- Since we are concentrating on the relationship between project management and systems engineering, we are necessarily focusing on the acquisition phase of the system life cycle; we address all phases and stages for completeness, but our emphasis is on the role of systems engineering in the projects that bring a system into being.

After reviewing the solid overview provided in this book, readers may seek more detail from the wide range of references provided at the end of each chapter or follow up with the two sibling books in this Artech House set: *Applied Systems Engineering* and *Requirements Practice in Conceptual Design: A Pracital Guide.* The overview here recognizes that the reader may come from any project-related discipline and with one of a wide variety of experience levels. *Applied Systems Engineering* provides a much deeper understanding of the body of knowledge of systems engineering across the life cycle. Subsequently, *Requirements Practice in Conceptual Design: A Pracital Guide* more thoroughly details the acquisition phase and particularly conceptual design, since the development of an appropriate set of system requirements is the most important task in system development.

Revision Questions

1. Briefly describe project management and systems engineering and their relationship.

2. Briefly define a system.
3. What is meant by an SOI?
4. What is meant by the terms operating environment and WSOI?
5. List the major resources that make up a capability system in the context of systems engineering.
6. Briefly outline how and why a system is described hierarchically (both logically and physically).
7. Briefly describe the differences between a subsystem, a system, and a system-of-systems.
8. Briefly describe the difference between the problem domain and the solution domain.
9. Briefly describe what is meant by a systems approach and what is meant by a soft systems approach and a hard systems approach.
10. Draw a diagram of and briefly describe the four broad phases of a generic system life cycle (preacquisition phase, acquisition phase, utilization phase, and retirement phase).
11. Identify the major responsible parties [enterprise management, business management, project management, systems engineering, the developer (contractor), and users/support] for the major phases of the life cycle.
12. Briefly describe the acquisition phase and the utilization phase of the system life cycle (as proposed by Blanchard and Fabrycky and used in this text), and explain briefly the activities that occur within each and the major artifacts and reviews associated with each.
13. List and briefly describe the following key themes of systems engineering: the top-down approach: requirements engineering, the focus on life cycle, system optimization and balance, the integration of disciplines and specializations, and management.
14. Briefly describe the three main categories of T&E: DT&E, AT&E, and OT&E.
15. Describe the benefits of using the systems engineering discipline during the development of a system.
16. Describe the analysis-synthesis-evaluation process that is applied iteratively throughout the system life cycle and explain each of the components.
17. Provide a brief description of the relationship between project management and systems engineering.

Endnotes

[1] Project Management Institute Standards Committee, *A Guide to the Project Management Body of Knowledge*, Upper Darby, PA: Project Management Institute, Version 6, 2017.

[2] Project Management Institute Standards Committee, *A Guide to the Project Management Body of Knowledge*, Upper Darby, PA: Project Management Institute, Version 6, 2017.

[3] Project Management Institute Standards Committee, *A Guide to the Project Management Body of Knowledge*, Upper Darby, PA: Project Management Institute, Version 6, 2017.

[4] Further reading on systems engineering can be obtained from the following major sources:

Aslaksen E. W., *Designing Complex Systems: Foundations of Design in the Functional Domain*, Boca Raton, FL: CRC Press, 2008.

Aslaksen E. W., *The Systems Concept and Its Application to Engineering*, New York: Springer, 2013.

Aslaksen, E., and R. Belcher, *Systems Engineering*, Upper Saddle River, NJ: Prentice-Hall, 1992.

Beam, W., *Systems Engineering: Architecture and Design*, New York: McGraw-Hill Book Co., 1990.

Blanchard, B. S., and J. E. Blyler, *Systems Engineering Management*, Hoboken, NJ: John Wiley & Sons, Inc, 2016.

Blanchard, B., and W. Fabrycky, *Systems Engineering and Analysis*, Upper Saddle River, NJ: Prentice-Hall, 2011.

Boardman, J., *Systems Engineering: An Introduction*, Upper Saddle River, NJ: Prentice-Hall, 1990.

Buede, D. M., and W. D. Miller, *The Engineering Design of Systems: Models and Methods*, New York: John Wiley & Sons, 2016.

Chestnut, H., *Systems Engineering Tools*, New York: John Wiley & Sons, 1965.

Chestnut, H., *Systems Engineering Methods*, New York: John Wiley & Sons, 1967.

Defense Systems Management College (DSMC), *Systems Engineering Management Guide*, Washington, Washington, DC: US Government Printing Office, 1990.

Dickerson, C. E., and D. N. Marvis, *Architecture and Principles of Systems Engineering*, Baco Raton, FL: CRC Press, 2010.

EIA-632, *Processes for Engineering a System*, Washington, DC: Electronic Industries Association (EIA), 2003.

EIA/IS-632, *EIA Interim Standard—Systems Engineering*, Washington, DC: Electronic Industries Association, 1994.

De Weck, O. L., D. Roos, and C. L., Magee, *Engineering Systems*, Cambridge, MA: MIT Press, 2011.

Dori, D., *Model-Based Systems Engineering with OPM and SysML*, New York: Springer, 2016.

Douglass, B. P., *Agile Systems Engineering*, Morgan Kaufmann, 2015.

Eisner, H., *Essentials of Project and Systems Engineering Management*, New York: John Wiley & Sons, 2008.

Faulconbridge, R. I., and M. J. Ryan, *Applied Systems Engineering*, Norwood, MA: Artech House, 2026.

Forrester, J., *Principles of Systems*, Cambridge, MA: The MIT Press, 1968.

Gangl, M., et al., *Modern Methods of Systems Engineering: With an Introduction to Pattern and Model Based Methods,* Createspace Independent Publishing Platform, 2011.

Grady, J.O., *Systems Integration,* Boca Raton, FL: CRC Press, 1994.

Grady, J. O., *System Engineering Planning and Enterprise Identity,* Boca Raton, FL: CRC Press, 2010.

Haberfellner, R., et al., *Systems Engineering: Fundamentals and Applications,* Cham, Switzerland: Birkhauser, 2019.

Hall, A., *A Methodology for Systems Engineering,* Princeton, NJ. D. Van Nostrand Co., Ltd., 1962.

Hitchins, D., *Putting Systems to Work,* Chichester, England: John Wiley & Sons, 1992.

Heap, G., J. Stanway, and A. Windsor, *A Structured Approach to Systems Development,* London: McGraw-Hill Book Company, 1992.

Hitchins, D., *Advanced Systems Thinking, Engineering, and Management,* Norwood, MA: Artech House, 2003.

Hitchins, D., *Systems Engineering: A 21st Century Systems Methodology,* Chichester, England: John Wiley & Sons, 2007.

Holt, J., *Systems Engineering Demystified: A Practitioner's Handbook for Developing Complex Systems Using a Model-based Approach,* Birmingham, UK: Packt Publishing, 2021.

Holt, J., and S. Perry, *SysML for Systems Engineering: A Model-Based Approach,* Institution of Engineering and Technology, 2013.

Hubka, V., and W. Eder, *Theory of Technical Systems,* Berlin: Springer-Verlag, 1988.

Hunger, J., *Engineering the System Solution: A Practical Guide to Developing Systems,* Upper Saddle River, NJ: Prentice-Hall, 1995.

IEEE-STD-1220, *IEEE Standard for Application and Management of the Systems Engineering Process,* New York IEEE Computer Society, 2005.

INCOSE, *Systems Engineering Handbook: A Guide for System Life Cycle Processes and Activities,* Walden, D. D, et al. (eds.), INCOSE-TP-2003-002-05 2023, Hoboken, NJ: John Wiley & Sons, 2023.

INCOSE, *Journal of the International Council on Systems Engineering,* Seattle, WA: International Council on Systems Engineering.

INCOSE, *Proceeding of Annual Conference,* Seattle, WA: International Council on Systems Engineering.

ISO/IEC/IEEE 15288, *Systems and Software Engineering—System Life Cycle Processes,* 2023.

Jackson, S., *Systems Engineering for Commercial Aircraft,* Surrey, England, Ashgate, 2015.

Jamshidi, M., *Systems of Systems Engineering: Principles and Practice,* Poca Raton, FL: CRC Press, 2009.

Jenney, J., et al., *Modern Methods of Systems Engineering,* CreateSpace Independent Publishing Platform, 2011.

Kamrani, A. K., and M. Azimi, *Systems Engineering Tools and Methods,* Boca Raton, FL: CRC Press, 2011.

Khisty, C. J., and J. Mohammadi, *Fundamentals of Systems Engineering with Economics, Probability, and Statistics,* Upper Saddle River, NJ: Prentice-Hall, 2012.

Kossiakoff, A., et al., *Systems Engineering Principles and Practice,* Hoboken, NJ: John Wiley & Sons, 2020.

Lacy, J., *Systems Engineering Management: Achieving Total Quality,* New York: McGraw-Hill Book Co., 1992.

Larson, W., et al., *Applied Space Systems Engineering,* New York: McGraw-Hill, 2009.

Machol, R., *System Engineering Handbook*, New York: McGraw-Hill, 1965.

Maier, M., and E. Rechtin, *The Art of Systems Architecting,* Boca Raton, FL: CRC Press, 2009.

Martin, J., *Systems Engineering Guidebook: A Process for Developing Systems and Products,* Boca Raton, FL: CRC Press, 1997.

MIL-STD-499B, *Military Standard—Systems Engineering—Draft,* Washington, DC: U.S. Department of Defense, 1994.

Micouin, P., *Model-Based Systems Engineering: Fundamentals and Methods,* London, John Wiley & Sons, 2014.

Parnell, G. S., P. J. Driscoll, and D. L. Henderson (eds.), *Decision Making in Systems Engineering and Management* Hoboken, NJ: John Wiley & Sons, 2008.

Rechtin, E., *Systems Architecting: Creating and Building Complex Systems,* Upper Saddle River, NJ: Prentice-Hall, 1991.

Reilly, N., *Successful Systems Engineering for Engineers and Managers,* New York: Van Nostrand Reinhold, 1993.

Ryan, M. J., *Requirements Practice in Conceptual Design: A Practical Guide,* Norwood, MA: Artech House, 2026.

Sage, A., *Decision Support Systems Engineering,* New York: John Wiley & Sons, 1991.

Sage, A., *Systems Engineering,* New York: John Wiley & Sons, 1992.

Sage, A., *Systems Management for Information Technology and Software Engineering,* New York: John Wiley & Sons, 1995.

Sage, A., and J. Armstrong, *Introduction to Systems Engineering,* New York: John Wiley & Sons, 2000.

Sage, A., and W. Rouse (eds.), *Handbook of Systems Engineering and Management,* New York: John Wiley & Sons, 2009.

Shishko, R., *ASA Systems Engineering Handbook,* Washington, DC: NASA, 1995.

Stevens, R., et al., *Systems Engineering: Coping with Complexity,* Hertfordshire, England: Prentice Hall Europe, 1998.

Sydenham, P. H., *Systems Approach to Engineering Design,* Norwood, MA: Artech House, 2004.

Thome, B. (ed.), *Systems Engineering: Principles and Practice of Computer-Based Systems Engineering,* New York: John Wiley & Sons, 1993.

Truxal, J., *Introductory Systems Engineering Management* New York: McGraw-Hill Book Co., 1972.

Wasson, C., *System Engineering Analysis, Design, and Development: Concepts, Principles, and Practices,* Hoboken, NJ: John Wiley & Sons, 2015.

Westerman, H., *Systems Engineering Principles and Practice,* Norwood, MA: Artech House, 2001.

Wheatcraft, L. S., M. J. Ryan, and T. E. Katz, *Needs and Requirements Manual: Needs, Requirements, Verification, Validation Across the Life Cycle,* Hoboken, NJ: John Wiley & Sons, 2024.

Wymore, A., *Model-Based Systems Engineering,* Boca Raton, FL: CRC Press, 1993.

[5] INCOSE, *Systems Engineering Handbook: A Guide for System Life Cycle Processes and Activities,* Walden D. D, et al. (eds.), INCOSE-TP-2003-002-05, Hoboken, NJ: John Wiley & Sons, 2023.

[6] ISO/IEC/IEEE 15288, *Systems and Software Engineering—System Life Cycle Processes,* 2023.

[7] Flood, R. L., and E. R. Carson, *Dealing With Complexity: An Introduction to the Theory and Application of Systems Science,* New York: Plenum Press, 1993.

[8] In the US Department of Defense, the acronym DOTMLPF-P refers to the capability system elements of doctrine, organization, training, material, leadership and education, personnel, and facilities—policy. [US Department of Defense, *Manual for the Operation of the Joint Capabilities Integration and Development System,* CJCSI 5123.01H, October 2021]. In Australia, the capability system is considered to comprise fundamental inputs to capability (FIC): command and management, organization, collective training, major systems, personnel, facilities and training areas, supplies, support, industry and data [Australian Government, Defence, *The One Defence Capability System (ODCS) Manual,* 2025]. In the United Kingdom, the defense lines of development (DLOD) refer to concepts and doctrine, equipment, information, infrastructure, interoperability, logistics, organization, personnel, and training [UK Ministry of Defence, *Acquisition Operating Framework,* version 1.1.13, May 2011]. In Canada, the acronym PRICIE refers to personnel, research and development, infrastructure, concepts and doctrine, information technology, and equipment (or, more completely, personnel, leadership, and individual training; research and development and operational research; infrastructure, environment, and organization; concepts, doctrine, and collective training; information management and technology; and equipment and support) [Canadian Department of National Defence, *Designing Canada's Army of Tomorrow: A Land Operation's 2021 Publication,* 2011]. In NATO, the acronym DOTMLPFI refers to doctrine, organization, training, material, leadership, personnel, facilities, and interoperability [NATO Standard ADatP-34, *NATO Interoperability Standards and Profile (NISP),* 18 August 2023].

[9] ISO 55000, *Asset Management — Vocabulary, Overview, and Principles,* 2024. ISO 55000 lists the following elements of an asset: people (competence, roles, leadership); processes (management systems, planning); information (data, documentation); physical assets; financial resources; support systems (IT, facilities, infrastructure).

[10] ISO/IEC/IEEE 15288, *Systems and Software Engineering—System Life Cycle Processes,* 2023.

[11] IEEE-STD-1220, *IEEE Standard for Application and Management of the Systems Engineering Process,* New York: IEEE Computer Society, 2005.

[12] ISO/IEC/IEEE 15288, *Systems and Software Engineering—System Life Cycle Processes,* 2023.

[13] ISO/IEC/IEEE 21839, *Systems and Software Engineering — System of Systems,* 2019.

[14] ISO/IEC/IEEE 15288, *Systems and Software Engineering—System Life Cycle Processes,* 2023.

[15] INCOSE, *Systems Engineering Handbook: A Guide for System Life Cycle Processes and Activities,* Walden D. D, et al. (eds.), INCOSE-TP-2003-002-05, Hoboken, NJ: John Wiley & Sons, 2023.

[16] Blanchard, B., and W. Fabrycky, *Systems Engineering and Analysis,* Upper Saddle River, NJ: Prentice-Hall, 2011.

[17] MIL-STD-499B, Military Standard—*Systems Engineering—Draft*, Washington, DC: US Department of Defense, 1994.

[18] Blanchard, B., and W. Fabrycky, *Systems Engineering and Analysis,* Upper Saddle River, NJ: Prentice-Hall, 1998.

[19] ISO/IEC/IEEE 15288, *Systems and Software Engineering—System Life Cycle Processes,* 2023.

[20] ISO/IEC/IEEE 15288, *Systems and Software Engineering—System Life Cycle Processes*, 2023.

[21] ISO/IEC/IEEE 29148, *Systems and Software Engineering—Life Cycle Processes—Requirements Engineering*, 2018.

[22] ANSI/EIA-632, *Processes for Engineering a System,* Washington, DC: Electronic Industries Association (EIA), 2003.

[23] MIL-STD-499B, *Military Standard—Systems Engineering—Draft,* Washington, DC: US Department of Defense, 1994.

[24] EIA-632, *Processes for Engineering a System,* Washington, DC: Electronic Industries Association (EIA), 2003.

[25] MIL-STD-499B, *Military Standard—Systems Engineering—Draft*, Washington, DC: US Department of Defense, 1994.

[26] EIA-632, *Processes for Engineering a System,* Washington, DC: Electronic Industries Association (EIA), 2003.

[27] INCOSE Fellows, *Systems Engineering and Systems Definitions,* 2018.

[28] INCOSE, *Systems Engineering Handbook: A Guide for System Life Cycle Processes and Activities*, Walden D. D., et al. (eds.), INCOSE-TP-2003-002-05, Hoboken, NJ: John Wiley & Sons, 2023.

[29] Defense Systems Management College (DSMC), *Systems Engineering Management Guide*, Washington, DC: US Government Printing Office, 1990.

[30] IEEE-STD-1220, *IEEE Standard for Application and Management of the Systems Engineering Process,* New York: IEEE Computer Society, 2005.

[31] EIA/IS-632, *EIA Interim Standard—Systems Engineering,* Washington DC: Electronic Industries Association, 1994.

[32] SECMM-95-01, *Systems Engineering Capability Maturity Model*, Version 1.1, Carnegie Mellon University, Pittsburgh, PA: Software Engineering Institute, 1995.

[33] Lake, J., "Unraveling the Systems Engineering Lexicon," *Proceedings of the INCOSE Symposium*, 1996.

[34] EIA-632, *Processes for Engineering a System,* Washington, DC: Electronic Industries Association (EIA), 2003.

[35] Wheatcraft, L. S., M. J. Ryan, and T. E. Katz, *Needs and Requirements Manual: Needs, Requirements, Verification, Validation Across the Life Cycle,* Hoboken, NJ: John Wiley & Sons, 2024.

[36] Boehm, B. W., "Verifying and Validating Software Requirements and Design Specifications," *IEEE Software,* 1984.

[37] Further details on these models can be found in, among others:

Blanchard, B., and W. Fabrycky, *Systems Engineering and Analysis,* Upper Saddle River, NJ: Prentice-Hall, 2011.

Martin, J., *Systems Engineering Guidebook: A Process for Developing Systems and Products,* Boca Raton, FL: CRC Press, 1997.

Young, R., *The Requirements Engineering Handbook,* Norwood, MA: Artech House, 2004, pp. 73–78.

[38] Forsberg, K., and H. Mooz, *Systems Engineering Overview,* Cupertino, CA: Center for Systems Management, 1996.

[39] Forsberg, K., and H. Mooz, "Application of the 'VEE' to Incremental and Evolutionary Development," *Proceedings of the NCOSE Conference,* St. Louis, MO, July 1995.

[40] EIA-632, *Processes for Engineering a System,* Washington, DC: Electronic Industries Association (EIA), 2003.

[41] CMU/SEI-94-HB-04, *A Systems Capability Maturity Model,* Version 1.0, Pittsburgh, PA: Software Engineering Institute, Carnegie Mellon University, 1994.

[42] Meier, S., "Best Project Management and Systems Engineering Practices in the Preacquisition Phase for Federal Intelligence and Defense Agencies," *Project Management Journal,* March 2008, pp. 59–71.

[43] Blanchard, B., and W. Fabrycky, *Systems Engineering and Analysis,* Upper Saddle River, NJ: Prentice-Hall, 1998.

[44] Project Management Institute Standards Committee, *A Guide to the Project Management Body of Knowledge,* Upper Darby, PA: Project Management Institute, Version 6, 2017.

[45] Washizaki, H., (ed.), *Guide to the Software Engineering Body of Knowledge (SWEBOK Guide),* Version 4.0, IEEE Computer Society, 2024; www.swebok.org.

[46] Faulconbridge, R. I., *Systems Engineering Body of Knowledge,* Canberra, Australia: Magpie Applied Technology, 2000.

[47] Faulconbridge, R. I., "A Systems Engineering Framework," *Journal of Battlefield Technology,* Vol. 3, No. 2, July 2000.

[48] MIL-STD-499B, *Military Standard—Systems Engineering—Draft,* Washington, DC: US Department of Defense, 1994.

[49] ANSI/EIA-632, *Processes for Engineering a System,* Washington, DC: Electronic Industries Association (EIA), 2003.

[50] IEEE-STD-1220, *IEEE Standard for Application and Management of the Systems Engineering Process,* New York: IEEE Computer Society, 2005.

[51] EIA-632, *Processes for Engineering a System,* Washington, DC: Electronic Industries Association (EIA), 2003.

[52] ISO/IEC/IEEE 15288, *Systems and Software Engineering—System Life Cycle Processes,* 2023.

[53] Kauffman, D., "Project Management and Systems Engineering: Where the Professions Intersect—Generate Synergy Not Conflict," *Proceedings of the Eighth Annual International Symposium,* INCOSE, 1998.

2

Conceptual Design

2.1 Introduction

The systems engineering processes begin with the identification by the business of a problem to be solvaed or an opportunity to be exploited. For simplicity we will, without presuming to narrow the discussion, focus on the solution of a problem and refer to that solution as the need for a new or improved system. The principal aims of conceptual design, therefore, are to articulate the mission for that system; to analyze, integrate, and record the business and stakeholder needs flowing from the mission; to transform those into an integrated set of needs and requirements; and to complete a functional design of the system as represented in the SyRS. The major product of conceptual design is the FBL, which provides, inter alia, a system-level functional architecture that is the basis for subsequent lower level design in the next stages, beginning with preliminary design.

Conceptual design is normally the domain of the customer/acquirer/tasking-activity. While other organizations may be involved in the functional design of the system, the business manager level of the customer must determine what the system needs to do and how well it needs to do it. Conceptual design is perhaps the most critical of all life cycle activities, because it is responsible for the expansion of a short statement of need in the form of a mission statement into a functional architecture that may be thousands of statements long.

This chapter introduces the major tasks undertaken during conceptual design. We begin by defining relevant terms and artifacts.

2.1.1 Concepts, Needs, and Requirements

When describing a system, we make a distinction between concepts, needs, and requirements. Concepts are typically narrative descriptions of ways in which the organization and its entities expect to manage, acquire, develop, deploy, operate, support, and retire the capability offered by the system. Needs are formal representations of expectations for an entity stated in the language and the perspective of stakeholders. Needs are then formally transformed into requirements, which are formal representations (model elements or textual statements) that communicate what an entity must do and how well it must do it to realize the intent of the needs. There may be more than one need for any given concept and more than one requirement defined for any one need.

Concepts, needs, and requirements exist at a number of levels corresponding to four major views, described as follows:

- An enterprise view, in which enterprise leadership sets the enterprise strategies and concepts of operations;
- A strategic view, in which business management derives business concepts and needs and formalizes its requirements;
- An operational view, in which stakeholders define their concepts, needs and requirements;
- A systems view, in which system designers define the system, first in functional and then in physical terms.

Because terms such as system and subsystem (system element) are level-specific—remember we are following a systems approach using a systems paradigm as discussed in Chapter 1—we use the generic term entity, which is "… a single thing to which a need or requirement refers: an enterprise, business unit, project, system, or system element (which could be a product, process, human, or organization)" [1].

The enterprise, strategic, and operational views are in the problem domain; the system view is in the solution domain. As discussed briefly in Chapter 1, the problem domain is the responsibility of those who have ownership of the problem to be solved (that is, those who are accountable for its solution), so the descriptions of the system are predominantly in the language of the customer, focusing on what they need the system to do, how well it should be done, and why. Recall from Chapter 1 that these descriptions are called functional (or often logical) descriptions. On the other hand, the solution domain is the responsibility of the developer implementing the solution, most commonly a contractor, so the descriptions of the system in that domain are predominantly in engineering and physical terms, focusing on how the problem is to be solved—that is, what it will look like and how it will operate once it has been implemented. As discussed in Chapter 1, these latter descriptions are called physical descriptions. In many large organizations, there is a soft boundary between the problem domain and the solution domain in that the customer prepares a draft SyRS that is taken to tender, responded to by tenderers, negotiated with the successful tenderer (the contractor), and then revised to be the SyRS against which the selected solution will be developed by that contractor. Section 2.5 discusses this process in more detail.

Although they coexist at each level of consideration, the concepts, needs, and requirements, described as follows, are different [2, 3]:

- A concept is a written or graphic representation that concisely expresses how an entity will satisfy the problem it was defined to address. A number of concepts can apply to each entity, so it is useful to develop a minimal set of concepts that define the expectations of all entities. It is also useful to provide some structure to the set of concepts. LCs include operations concepts, acquisition concepts, deployment concepts, support concepts, and retirement concepts.
- A need is the result of a formal transformation of one or more concepts into an agreed-upon expectation for an entity to perform some function or possess some quality (within specified constraints with acceptable risk). Needs are generally capabilities stated as business needs in the language of business management, as stakeholder needs in the language of stakeholders at the business operations level, or as needs of the system. There will generally be a number of needs defined for any one concept.
- A requirement is the result of a formal transformation of one or more needs or parent requirements into an agreed-to obligation for an entity to perform some function or possess some quality (within specified constraints with acceptable risk). Requirements are formal statements that are structured and can be validated. Requirements are elaborated from needs through a process of requirements analysis (also called business analysis or mission analysis [4] at the highest level of application). There can be more than one requirement defined for any one need.

ISO/IEC/IEEE 15288 and the INCOSE *Systems Engineering Handbook* define that, at the highest level, there is an enterprise view in which enterprise leadership sets business vision, and strategic goals and objectives in the form of a concept of operations (CONOPS), which is also called a strategic business plan (SBP). At the next two levels, there are the dual views: at the business management level there are business needs [5] and business requirements [6]; and at the business operations level there are stakeholder needs [7] and stakeholder requirements [8]. The transformation of needs into requirements occurs at each level. Here we are interested in the three major upper-level processes, described as follows:

- *BNR definition process:* At the strategic level, business management begins the process by providing the guidance contained in the CONOPS, from which they define business needs, largely in the form of preliminary LCs (PLCs) which capture the preliminary acquisition concept, preliminary operational concept (OpsCon) [9], preliminary deployment concept, preliminary support concept, and preliminary retirement concept. Business needs are elaborated and formalized into business requirements, which are recorded in the business requirements specification (BRS) [10], which is also called the business requirement document [11].
- *SNR definition process:* Using the ConOps, the preliminary OpsCon, and the other PLCs as guidance, stakeholders from the business operations level develop stakeholder needs—captured in the form of a refined OpsCon and other LCs—and transform them into a formal set of stakeholder requirements, which are recorded in the Stakeholder Requirement Specification (StRS) [12].
- *System requirements definition process:* The requirements in the StRS are then transformed by requirements engineers into system requirements, which are contained in the SyRS [13]. This artifact is also referred to as the solution requirement specification [14], or simply the system specification [15]. More than one SyRS may be derived from the one StRS. For example, as discussed in Chapter 1, an SyRS may be produced for each element of the capability described by the StRS (see Figure 2.5). Some organizations may also prepare individual LCs for each of a number of elements of capability that are developed to meet the business needs. The process continues for the system elements.

It should be noted that, in the past, almost every standard, textbook, and academic paper would have a slightly different spin on systems engineering terms, language, and processes, among other industry-specific descriptors. However, with the revision of ISO/IEC/IEEE 15288 in 2023 and the subsequent revision of the INCOSE *Systems Engineering Handbook* and the associated INCOSE manuals and guides (see Chapter 6 for a summary), the language is gradually beginning to converge. The description above conforms with ISO/IEC/IEEE 15288, the INCOSE *Systems Engineering Handbook,* and ISO/IEC/IEEE 29148 [16] with the exception that the business or mission analysis process in ISO/IEC/IEEE 15288 is called the BNR definition process in this text (the activities in the processes are consistent).

Additionally, for practical implementation, the INCOSE *Needs and Requirements Manual* (NRM) [17] combines what ISO/IEC/IEEE 15288 calls the BNR and SNR into an integrated set of needs, which is then transformed into design-input requirements (designated as the SyRS by ISO/IEC/IEEE 29148). Despite a preference for the structure of the INCOSE NRM, we conform in this text to ISO/IEC/IEEE 15288 and ISO/IEC/IEEE 29148 and the INCOSE *Systems Engineering Handbook,* because systems engineers are certified against those artifacts, and many organizations use them to provide the frameworks for their projects (especially those projects that are already in progress).

In any case, it is of little consequence because the reader can readily make the translation; the BNR and SNR described in ISO/IEC/IEEE 15288 are combined in the NRM's integrated set of needs, and the SyRS described in ISO/IEC/IEEE 29148 is called the design-input specification in the NRM to help visualize the relationship. Figure 2.1 shows both taxonomies mapped to the processes and artifacts used in this book. To assist throughout, we occasionally use both sets of terms to reinforce the duality.

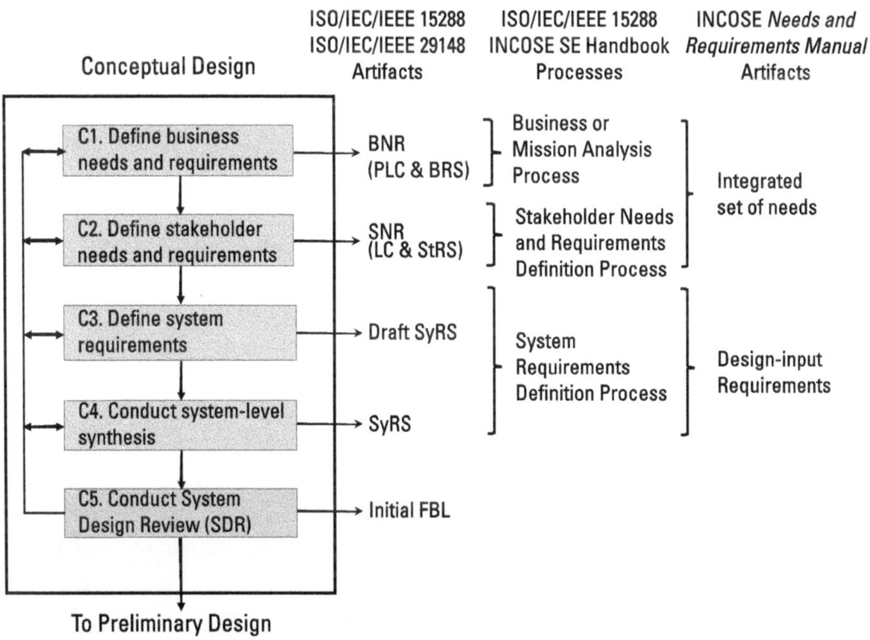

Figure 2.1 The conceptual design processes and needs and requirements artifacts mapped to those called for by ISO/IEC/IEEE 15288 and ISO/IEC/IEEE 29148, and the artifacts from INCOSE's *Needs and Requirements Manual.*

2.1.2 Conceptual Design Tasks

Figure 2.1 shows that the following five major tasks are conducted during conceptual design:

- Definition of the BNR in the PLCs and the BRS;
- Definition of SNR in the LCs and StRS;
- Definition of draft system requirements in the draft SyRS;
- System synthesis to finalize the SyRS with the preferred solution in mind;
- SDR to establish the initial FBL.

These five tasks are described in more detail in the following sections.

2.2 C1. Define Business Needs and Requirements

Before any work can commence on developing the system, the basic BNR must be clearly and completely articulated. It would appear obvious that an idea of what needs to be achieved must be clearly understood and articulated before any further work is undertaken. Yet, a surprising number of system developments commence without a clear and complete understanding of the fundamental needs and requirements. Not surprisingly, most of these developments founder or, at the very least, are introduced into service with poor levels of user acceptance.

Figure 2.2 summarizes the activities and steps involved with the definition of BNR resulting in the development of the PLCs and the BRS. Note once again that the process is iterative, allowing current understanding to improve the quality of previously generated considerations. The activities of Figure 2.2 are described in more detail in the following sections.

2.2.1 C1.1 Identify Major Stakeholders and Constraints

At the strategic level of the organization, business management begins the definition of BNR with the identification of the major stakeholders who are to be engaged to elicit business needs and requirements and to verify design artifacts as the design progresses. Business management must also identify the business, project, external, and design constraints that provide important context within which to begin consideration of the system requirements and that are the source of many other project and support requirements. The first step in developing the BNR (Figure 2.2, activity C1.1), therefore, requires the identification of the major stakeholders responsible for the business (the strategic) aspects of the new

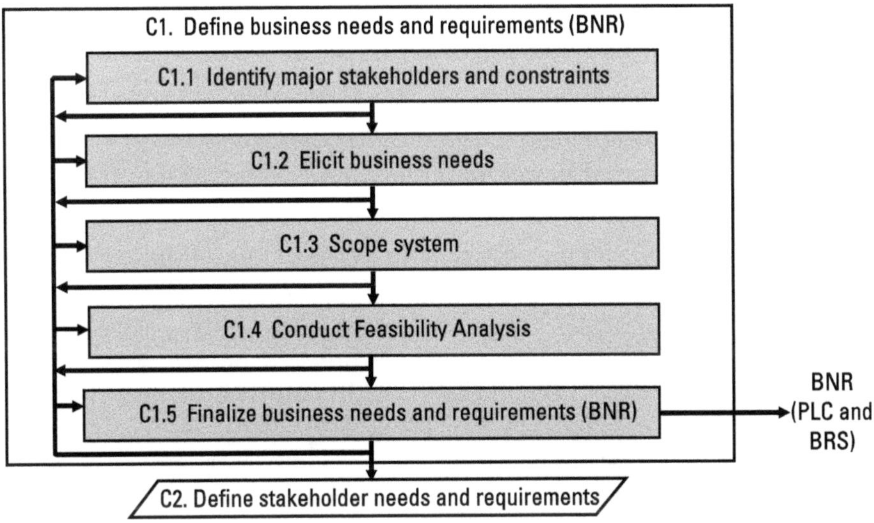

Figure 2.2 A summary of C1. Define business needs and requirements.

system as well as any constraints that the business wishes to place on the system or on its development.

2.2.1.1 Identify Major Stakeholders

Stakeholders will be responsible for writing and agreeing to the BNR prior to them being endorsed by the business. These stakeholders will also be called on during later activities to assist with the requirements analysis effort. In the first instance these stakeholders will be represented by their business units for the development of BNR. Lower level stakeholders make their major contribution during the development of SNR.

Example 2.1: Stakeholders for the Aircraft System

There are a number of stakeholders within ACME Air who provide a major source of information relating to the functional and performance needs of the aircraft system. Corporately, the business owners of the airline have a set of more commercial needs for the aircraft, such as fuel efficiency, passenger and freight capacity, and so on. Users, including aircrew and ground crew—personnel who will eventually operate the aircraft—will have significant input to the functional and performance needs. Marketing people dictate the features necessary in the aircraft system to make the aircraft marketable to passengers. Maintainers influence the design to ensure that the aircraft is maintainable within the time and cost constraints set by the business case. Accessibility, reliability, and maintainability will be high on the priority list of the maintenance representatives. Logistics and supportability issues such as the provision of spare parts are also likely to influence design decisions.

2.2.1.2 Identify Business, Project, External, and Design Constraints

Before focusing on the details of the desired system, it is essential that the business stakeholders identify the major constraints that are relevant to the system and its acquisition. Typically, constraints are considered in the general categories of business, project, external, and design.

Business constraints include management guidance and any organizational policies, procedures, standards, or guidelines that steer system development and procurement. These constraints can include partnering relationships with other companies and contracting policies, among others. Project constraints generally flow from business constraints and include the resource allocations to the project as well as any externally imposed deliverables and acquisition timeframes. Many companies have enterprise-wide standards for processes such as quality assurance and systems engineering, and these methodologies guide the way projects can operate. Additionally, the business may require the project to report progress in a particular way or to implement particular metrics, tools, and documentation/recording procedures.

In addition to business-imposed constraints, there are wider external constraints on system development that arise from the need for conformance to national and international laws and regulations, compliance with industry-wide standards, and ethical and legal considerations. Other external constraints include the need for interoperability and the capabilities required for interfacing to other systems. It is an important aspect of top-down design to understand these constraints before considering lower level system requirements.

Design constraints directly affect the way in which the system design can be conducted. Typical constraints include the state of the art of relevant technologies as well as extant methodologies and tools to assist in the design, development, construction, and production of the system. Such issues must be considered by business management who will address the risk associated with embracing new technologies, the impact of upgrading construction facilities, workforce reskilling issues, among other considerations.

Example 2.2: Example Constraints for the Aircraft System

An example of a business constraint for ACME Air would be a policy of only buying engines from a particular manufacturer to ensure that through-life support issues are simplified across the entire fleet. This constraint must be articulated early as it has a significant effect on the procurement.

Project constraints will include budget and schedule but also cash-flow constraints (availability of funding), location(s) of the work, preferred part-

nering arrangements, project methodology, project team size and location, and many more.

Additionally, bounds such as all-up mass may be a stated design constraint for an aircraft to ensure it meets weight-related restrictions at certain classes of airfield.

2.2.2 C1.2 Elicit Business Needs

Having established the context by identifying major stakeholders and constraints, the next activity is to elicit business needs (Figure 2.2, activity C1.2). The major activities are the development of the mission, goals, and objectives; PLCs; and preliminary validation criteria.

2.2.2.1 Develop Mission, Goals, and Objectives

Every project should begin with a concise statement of the mission, elaborated by statements of the upper-level goals and objectives. The mission statement should be stated in a single sentence, although it should have a word or phrase for every important aspect of the system. While business management often finds it difficult to state the mission in a single, short sentence, the project is invariably doomed to failure if the owners of the mission cannot articulate it succinctly at the outset. Interestingly, it will often take a small number of participants several hours to craft and agree to a mission statement; the process is useful, therefore, not only to start the design process, but also to serve as a mechanism for developing among stakeholders a common understanding of the system. That it could take several hours to agree to a mission statement should also act as a warning that the project is doomed to considerable confusion if the stakeholders continue, each with a differing view of what the system is to be.

The mission statement is then expanded and qualified by short declarative statements of the system goals and objectives. Goals are normally relatively broad statements, each of which spawns a number of more-specific objectives (although these are sometimes treated in reverse order and objectives are considered to lead to goals).

Example 2.3 Mission Statement for the Aircraft System

For the ACME Air example, the mission statement for our aircraft system might be the following:

> *... to provide a medium-sized aircraft that can provide class-leading comfort to passengers between Class X airfields on domestic and international routes.*

In this example, the goals and objectives would then elaborate on such matters as what is meant by "class-leading" and "medium-range," how many passengers are to be carried, as well as operational issues such as crewing and maintenance.

Example 2.4: Mission and Goals for the Domestic Security Alarm System

For the ACME Domestic Security Alarm, the mission statement might be the following:

> *...to provide a market-leading alarm system that can deter, detect, classify, and report unauthorized entry to a residence in order that the resident is aware of the residence's state of security.*

This mission statement could then be supported by the following goals, which can be decomposed from the mission:

- *Deter unauthorized entry;*
- *Detect unauthorized entry;*
- *Classify unauthorized entry;*
- *Report unauthorized entry;*
- *Provide market-leading features.*

The needs and requirements form a hierarchy, first in a functional hierarchy in the BRS, the StRS, and the SyRS, and then in a physical hierarchy in the subsystems and configuration items in the development specifications. Traceability is assisted through the use of hierarchical numbering.

The draft mission, goals, and objectives may be confirmed by communicating them to the broader community of stakeholders, not just those immediately involved in their development. Once confirmed, the mission, goals, and objectives are finalized into a formal set of need statements that provide a complete, balanced description of the system, covering all aspects of operation, support, and maintenance (albeit at a high level of abstractions). The INCOSE NRM calls this set the set of integrated needs [18]. The set is the starting point for the system design to which every subsequent requirement must be traceable.

2.2.2.2 Define PLCs

Once the mission, goals, and objectives have been articulated, the top-down process is continued through an examination of operational scenarios that are relevant to the system. The examination begins with a description of the general operational environment for the system to identify all the environmental factors that may influence the operation of the system. Specific operational scenarios are then described in stakeholder language to depict the full range of circumstances under which the system is required to operate. It is not necessary to describe every possible scenario, but all types of operation must be represented. Scenarios also need to represent all stakeholder perspectives. For example, turning an aircraft around requires different activities from aircrew and groundcrew; both have valid views that spawn different subsets of needs and requirements.

These scenarios, associated with what are called use cases or vignettes, provide valuable guidance to the system designers and form the basis of major events in the acquisition phase such as validation of the system as it is introduced into service. Despite any more detailed technical verification and validation procedures, the system's fitness for purpose is fundamentally related to its ability to perform in accordance with the operational scenarios defined at this stage.

In many cases it is also useful to define the various modes of operation for the system products under development. Designers need to understand whether the system is to operate in different modes—even if it is as simple as the difference between the fully operational mode or the training mode. Complex systems may have their needs and requirements stated in a number of modes. For example, a modern fighter aircraft may have modes defined for air-to-air combat, ground attack, reconnaissance, naval operations, nontactical flights, and air-show deployments, among others. Each mode must be associated with the particular conditions (i.e., mission, operational, environmental, or configurational) that define it.

In our aircraft example, a number of modes may be defined for international and domestic operation including taxi, takeoff, cruise, approach, landing, and turnaround. Modes may also be defined for maintenance and for administrative movement of the aircraft.

Having defined a preliminary operational concept, business management must give some guidance on the other LCs related to the acquisition, deployment, support, and retirement of the system. While the systems engineering procedures that follow will ensure a life cycle focus, it is important that there is an early business focus on the major cost drivers that will impact on all aspects of the system throughout its life. Additionally, there are a number of life cycle-related trade-offs at a business-case level that need to be concluded to inform subsequent considerations (perhaps providing constraints).

Example 2.5: Life Cycle-Related Trade-Offs

PLCs are continuously tested and adjusted, if necessary, throughout subsequent design as the system is understood in more detail. Consider the example of a need for an inexpensive dashboard-mounted Global Positioning System (GPS) for a large trucking company. Since the item is to be procured in large numbers at a low cost, the business case may originally deem that it would not be cost-effective to implement a repair system for defective items, which could be discarded when broken. That preliminary concept may well be adjusted during subsequent more detailed consideration to recognize that there would still be a need to inspect "broken" items before being discarded to ensure that they were indeed inoperative, not just subject to problems such as finger faults or sticking buttons that could be rectified simply. If technical resources are to be applied to inspection, perhaps the additional cost of slightly more detailed training may allow some items to be repaired rather than discarded, resulting in a reduction in total cost of ownership (TCO).

It should be noted that the LCs must be considered in the context of the capability required to meet operational scenarios (which is why the preliminary operational scenarios and concepts are developed before the other PLCs). For example, the support aspects of a system must not be optimized in their own right without consideration of the operational use of the system; in other words, the lowest-cost through-life support for an aircraft is achieved by minimizing fuel consumption and reducing wear and tear that requires maintenance—in short, by not operating it all. Clearly, therefore, through-life support must be optimized based on the desirable operational scenarios, not independent of them. Importantly, however, the business must also recognize that through-life support optimized for the operational scenarios may not necessarily be optimal for any other scenarios should the operational context be changed in the future.

Before looking briefly at the content of each of the PLCs, it should be noted that those concepts are indeed preliminary and will be adjusted over time as more is understood about the system and the life cycle issues, The value of the preliminary investigations, however, is to understand the options available in each concept so that they can be iterated and adjusted to be optimized as a set.

2.2.2.3 Define Preliminary Validation Criteria

An important top-level activity is the identification and definition of preliminary validation criteria. Broadly, validation criteria encompass any mechanism by which the acquirer will measure satisfaction with the products of the acquisition phase. Key criteria include performance in each operational scenario, safety, reliability, supportability, maintainability, ease of use, and time and cost to train, among many others.

For a measure to be relevant, it must be related to one or more requirements. One common way of doing this is to develop a matrix that relates measures to requirements. In fact, the use of relationship matrices is very common in systems engineering: We have traceability matrices that show inheritance of requirements; we have verification matrices to show how verification means/methods relate to individual requirements or groups of requirements; and we could have matrices to relate measures to requirements.

2.2.3 C1.3 Scope System

The next phase in successful requirements analysis necessitates an understanding of the scope of the system development effort (called scoping). This scoping process helps to establish a clear understanding of what the system is expected to do by soliciting, determining, and detailing design constraints, external interfaces, and system boundaries.

To assist with the scoping process, a tool called a context diagram may be used to illustrate the interrelationships of related systems, relevant regulatory environments, stakeholders, external systems, and any interfaces.

Example 2.6: Simple System Context Diagram for Our Domestic Security Alarm Example

Figure 2.3 illustrates a simple context diagram for the operational view of our domestic security alarm example; the acquisition and support (particularly maintenance) views would be shown in a separate diagram. In reality, of course, the diagram would be a little more detailed, but Figure 2.3 shows the main elements of a useful diagram. Additionally, even this operational

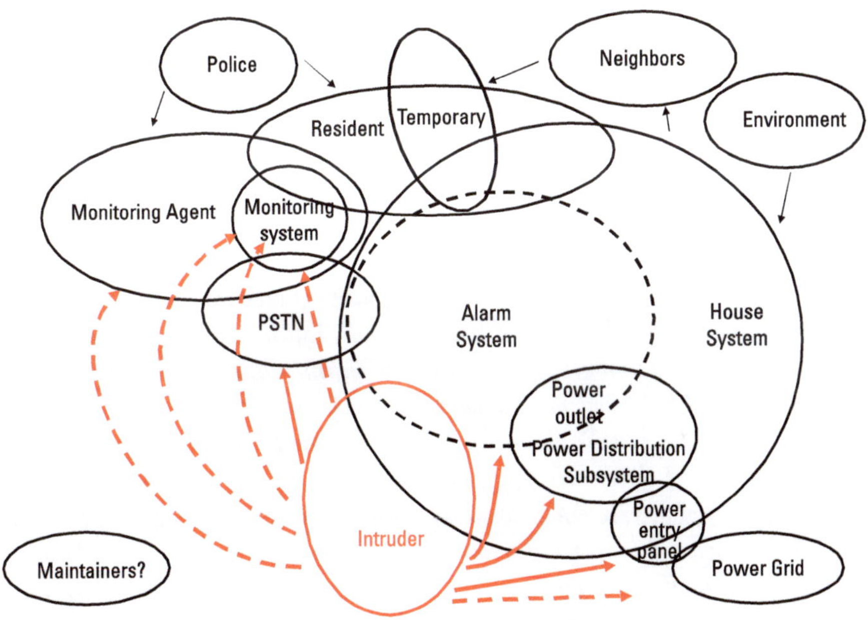

Figure 2.3 Simple context diagram for our domestic security alarm example.

diagram would be best split to show the normal state of operation (when the intruder is not present) and the exception case (when the resident is absent, but the intruder is present). The elements of the domestic security alarm and their operation are discussed as follows.

- Alarm: *It is obvious from the context diagram that the alarm is entirely contained within the larger house system since one goose egg is a subset of the other. It is also obvious from the way the alarm subset is drawn that the designers accept the prospect that the alarm might not cover 100% of the area covered by the house system (since the boundaries of the two are not coincident). So, in this case, the context diagram shows a relationship between two elements; one is totally subsumed by the other.*
- Resident: *The resident has a number of interfaces evident on the context diagram—shown as overlapping goose eggs. There are, of course, the physical interfaces in which the resident first enters the house and then enters the sphere of influence of the alarm. In this case, the context diagram shows an interface of the resident with the house and the interface of the resident with the alarm. More than that, however, it shows that the resident's interface with the alarm occurs after their interface with the house—that is, the resident has no relationship or interface with the alarm until they are inside the house. Also evident is the interface between the resident and the monitoring agent, which is shown as being independent of the alarm system. Further, the diagram includes the notion that there might be temporary residents (guests of the residents, for example) who will interface with the alarm (in the same way as a resident) for the period during which they are guests.*
- Intruder: *The intruder has a physical interface that is very similar to the resident (both are humans that may enter the house and then the alarm boundary by a number of means), so it can be concluded that the alarm must be able to distinguish between the two in some manner. Note that, in addition to the physical interface, the diagram also shows a number of influences that the intruder may have on the alarm and house systems; the solid lines show the influences of most interest in this design (disabling alarm sensors or power, disconnecting house power, disconnecting the phone line); and those that the threat analysis considers less likely for a domestic burglary (disabling the suburban power grid, logging across the phone line to mimic the alarm, hacking into the monitoring system, paying an accomplice inside the monitoring agent to disable or ignore the alarm, or destroying the monitoring*

agent). Part of the scoping activity, then, is to discard these latter possible actions by the intruder as beyond the scope of a domestic security alarm.

- Monitoring agent: *Note that there is an expectation in this design that there is a monitoring agent (one does not necessarily have to exist in an alarm system) and that the mandated interface is via the Public Switched Telephone Network (PSTN) (rather than a more modern version of alarm where a cellular data service might be incorporated). In this case the context diagram assumes the presence of an element (the monitoring agent) and the use of a technology (the PSTN); both assumptions are clearly obvious in Figure 2.3, which shows that the context diagram is a very useful communications device, since such assumptions should be surfaced and tested as early as possible. We test this assumption in the Example 2.7.*
- Power: *The diagram shows the expected interface of the alarm to the house power subsystem. (Note that since a power outlet is identified explicitly, we can infer that the alarm is not hard-wired into the power subsystem (perhaps by a licensed electrician) but instead is able to be plugged in by the resident so that the alarm can be transported between houses.*
- Police: *The context diagram illustrates that the police do not have any direct interface wtih this system (i.e., that they are not expected to respond when the alarm is triggered but would respond to any incident as a function of their normal policing duties).*
- Neighbors: *Again, neighbors are shown to have no particular role in the alarm system but are affected by it (perhaps by the noise it may make) and may influence the resident (perhaps by complaining about the noise).*
- Maintainers: *The diagram shows that there has not yet been a decision regarding the way in which maintenance is to be conducted. Such an arrangement, once known, will be able to be reflected in the diagram, or perhaps in a separate support context diagram since operations and maintenance are unlikely to be undertaken at the same time.*

The context diagram assists in capturing stakeholder needs and requirements, but it is also useful to assist in performing the necessary feasibility analyses and upper-level trade-offs at the business level. That is, the context diagram is a useful visual design artifact that allows stakeholders to visualize the system at an abstract level to support them to make decisions at that level that will then drive (constrain) the subsequent considerations.

Example 2.7: Example Upper-Level Trade-Offs Using the Context Diagram for Our Domestic Security Alarm Example

> *Even at this early stage, there are a number of upper-level trade-offs that can be conducted for our security alarm example. When developing the context diagram, the stakeholders included a monitoring agent connected to the alarm via the wired telephone system (the PSTN). Before progressing and transforming those implied needs into system requirements, there are some decisions to be made. For example, is the PSTN still relevant—after all, how many households have a fixed phone line these days? Would an Internet connection or a mobile phone interface be more relevant? If so, which one? Or both? Or should we remain technology agnostic at this early stage?*
>
> *Similarly, why is there a monitoring agent? Although the business originally requested a back-to-base option, that was an essential element of such an alarm in the past, but in a modern system the alarm could report directly to the resident when they are anywhere in the world (either across the internet, or by message via mobile phone, or both). Would the business like to reconsider its original need now that there is more detail available?*
>
> *The context diagram also assists in the definition of the system boundary, which is critical to the success of the fielded system. It is essential that the boundary is defined early in the acquisition phase so that it is clear which system elements are under the design control of the project, and which are outside the control. This is also particularly important to the project manager who is concerned with defining what is to be included in the system as well as what is to be excluded. In that sense, the context diagram is an essential tool in the project management activity of scope management.*
>
> *The boundary on the context diagram also assists in the identification and definition of interfaces required with external systems, as these will place considerable requirements on the system under development. While these external systems are not directly related to the project, the success of the fielded system is often determined by its ability to interface to its external environment. For example, while it is possible to build a perfectly functional aircraft without consideration of air traffic control regulations, the aircraft would be useless to the business because it would not be allowed to operate.*

2.2.4 C1.4 Conduct Feasibility Analysis

Once business management has defined their needs, the next major activities are related to the transformation of those needs into formal requirements, perhaps in the form of the BRS in accordance with ISO/IEC/IEEE 29148. This last step is really only useful if the business intends to use those requirements to engage a design partner to complete the design, especially if the engagement is via a formal contract. Otherwise, the process can continue with the business needs, which can then be combined with the stakeholder needs in what the INCOSE NRM [19] refers to as the integrated set of needs (as an alternative to the development of the BNR/SNR and BRS/StRS).

At this stage of conceptual design, business management has stated their needs in logical terms—that is, in terms of the problem domain (the management and operational environment of the business). For any problem domain, however, there is normally a range of potential classes of solutions within the solution domain. Each alternative solution class may represent a completely different type of project. The feasibility analysis is therefore an essential step to narrowing the solution domain so that the subsequent project can be managed effectively. If we continue without narrowing down the design into one of those classes, we will have great difficulty coping with the breadth of the design. Consequently, we next undertake a feasibility analysis, the aim of which is to select the desired solution class (which is then to be elaborated further by the stakeholders at the business operations level).

Business requirements are generally couched in business terms or capability terms and are elicited first so that we can understand what is required by the business of the deployed system. Requirements analysis elaborates these BNR into SNR and then goes on to transform SNR into system requirements. The idea of a feasibility analysis is to pause after identifying BNR to determine what possible system-level alternatives are available before going on to conduct detailed requirements analysis and eventually to establish the system requirements. In other words, there is most likely a variety of broad classes of systems that might meet the upper-level BNR, and it is often very useful to choose one of these before proceeding to determine stakeholder needs and requirements, or else the SyRS that would result would be far too general to be useful [either for subsequent design or, more particularly, as a centerpiece of the statement of work (SOW) for a contract]. Not only that, but the choice among solution classes is invariably a business choice and should therefore be made before time is wasted in elaborating classes of solution that the business would not wish to embrace.

Once the BNR have been articulated, the feasible system-level alternatives available to meet those needs can be investigated. Note that does not mean that the business must choose an alternative after feasibility analysis—they may not be able to. It is just that, if a feasible alternative is available, the remainder of the process will be much more concise and focused.

Feasible alternatives are considered in terms of available resources such as money, time, personnel, and materials. The following steps are completed during feasibility analysis:

- Identifying the possible system-level solutions capable of satisfying the BNR;
- Confirming the compliance of each solution class with the BNR and, if not completely compliant, noting the likely level of achievement;
- Evaluating the alternative solution classes in terms of feasibility, performance, effectiveness, technical and project risk, and other selected measures:
- Selecting the best of the alternative solution classes (ensuring that the options are narrowed down as much as is practicable at this stage).

A great deal of care should be taken during this stage of the project, as the feasibility analysis sets the direction for the rest of the system life cycle. To that end, the feasibility analysis should address the entire life cycle, not just the operational aspects of the system. (Recall Example 1.4 where the feasibility analysis for the RAAF F-111 could have had an impact on the system for more than 50 years.)

There may be an opportunity to examine the marketplace to identify existing systems that can be assessed to assist with feasibility analysis. Assessing operational systems may also assist with the refinement of the mission, goals, and objectives of the new system based on a realistic understanding of what is currently available. The feasibility of using existing systems to meet emerging needs may be assessed as a possible way of reducing the schedule and technical risk associated with the system development. The process is sometimes called concept demonstration or technical demonstration.

Example 2.8: Feasibility Analysis for Alternatives for Replacing the Aircraft System

ACME wants to service the short-to-medium domestic and international routes but doesn't have the airframes to do it. The feasible alternatives are listed as follows:

- *Leasing or buying an existing aircraft [off-the-shelf (OTS)];*
- *Leasing or buying and modify an existing aircraft (modified-OTS);*
- *Contracting to have a new aircraft developed;*
- *Outsourcing the operation to another airline;*
- *Not servicing these medium routes.*

There is therefore a wide range of possible solution classes for ACME in replacing its existing aircraft. Then, for each approach, there is still a range of subordinate alternative system solutions. Each of the alternatives is a completely different type of project; feasibility analysis is therefore an essential step in narrowing the solution domain so that we can manage the subsequent project. It is also an important step for the business to confirm for itself which type of solution it prefers.

The feasibility analysis narrows the solution domain to reduce the scope of the resultant requirements that are recorded in the BRS. It should also be noted that reducing the solution classes to one preferred class is a business management decision, and not one that should normally be made by stakeholders at the business operations level, nor by project management, nor by systems engineering.

Another possible result of the feasibility analysis could be determining that no preferred system solution exists (or at least exists at a sufficient level of maturity or within the organization's budget or schedule), and the remaining process may not continue because the business case cannot be met at the current time.

2.3 C2. Define Stakeholder Needs and Requirements

The endorsement of the BNR signifies that business management is confident that they have defined the problem domain in sufficient detail to communicate their needs and requirements to the stakeholders at the business operations level. These stakeholders then go on to develop their needs and requirements (the SNR) within the context of the needs and requirements and constraints of the BNR.

Figure 2.4 illustrates the activities involved with the process of defining SNR to subsequently result in the development of the LCs and the StRS. Note once again that the process is iterative, allowing current understanding to

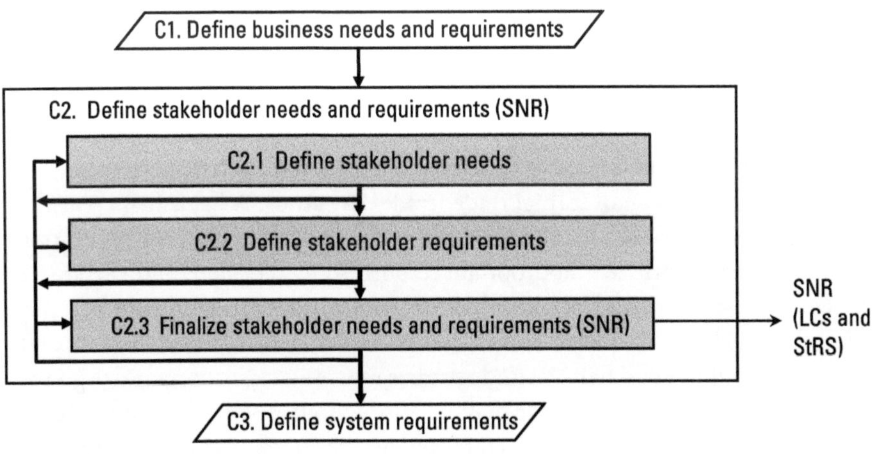

Figure 2.4 A summary of C2. Define stakeholder needs and requirements.

improve the quality of previously generated artifacts. Sections 2.3.1 and 2.3.2 describe the activities of Figure 2.4 in more detail.

2.3.1 C2.1 Define Stakeholder Needs

Once the BNR have been confirmed for the selected solution class, the nominated stakeholders at the business unit level can begin to articulate their needs and requirements. Although the major stakeholders were nominated by business management during the definition of BNR, the first act in the process of defining SNR is to refine those stakeholders by naming individuals within the major stakeholder groupings and adding lower level stakeholders at the business operations level.

Within the context set by the BNR, stakeholders at the business operations level articulate their needs through the refinement of the mission, goals, and objectives. The preliminary operational scenarios defined by business management are now defined in more detail, and other lower level scenarios (often called vignettes) are developed as the design progresses. Similarly, the other PLCs are developed further and fleshed out with details from the various stakeholders to be more complete concepts in the LCs.

Once complete, stakeholder needs are endorsed first by the stakeholders at the business unit/operations level and then by business management. Again, endorsement is most likely achieved through a major review.

2.3.2 C2.2 Define Stakeholders Requirements

Once the stakeholder needs have been endorsed, the process of transforming their needs into formal stakeholder requirements is undertaken. A hierarchical structure is a useful framework for refining the BRS into the detail of the StRS while also assisting in the maintenance of traceability; we will discuss the RBS in Section 2.4.1.

In a similar manner to the feasibility analysis that business management used to narrow down the alternative solution classes, stakeholders at the business operations (business unit) level will have a number of alternative ways in which their needs can be met. The selection between various options is best undertaken by some type of formal trade (trade-off) study. A list is first created of the trade studies that are required; related trade studies are grouped for efficiency; studies are prioritized; appropriate resources are allocated, and the studies are conducted. Of course, the conduct of trade studies did not begin here (the feasibility analysis conducted earlier can be considered as a system-level trade study) and will not end here, as trade studies represent an important design tool that is used to support decision making throughout system development and the remainder of the life cycle. Chapter 6 describes trade studies in more detail.

2.3.2.1 StRS

The development of the StRS is an essential step toward successful system development and is given a number of titles including, simply, stakeholder requirements or user requirements. A reader of the StRS should be able to understand completely the likely applications or missions for which the system is intended; the major operational characteristics to be exhibited by the system; the operational constraints that limit the design and development of the system; the external systems and interfaces with which the system under development must operate; the operational and support environment within which the system must exist; and the support concept to be employed to support the system and enable it to continue performing in accordance with customer expectations.

Although the StRS is a formal requirements artifact, it must be written in the language of the stakeholders to ensure that it is readable by all stakeholders. The StRS is not intended to be the FBL of the system and should therefore not be written using the formal specification language employed in the SyRS and later specifications. However, since it is critical that the StRS is comprehensive, readable, unambiguous, and subsequently easily traceable, the language must still be formal. In many cases the difference between the StRS and the SyRS is marked, because the user is operationally focused and utilizes very little technical language. In other cases, such as our aircraft system, the stakeholders are groups such as the air crew, ground crew, maintainers, and engineers, all of whom will use relatively technical language in their StRS statements. The difference is always, however, that StRS statements are high-level and made in more natural language, which is generally much easier to read and understand than the language of formal specifications.

Again, as for the business requirements, the StRS (as required by ISO/IEC/IEEE 29148) does not necessarily have to be prepared, and the BNR/SNR can be consolidated in the integrated set of needs (as suggested by the INCOSE NRM [20]). On the other hand, the business may wish to engage a design partner at this point, in which case they may formalize the StRS to be used as the basis of a partnering agreement.

Example 2.9 StRS Language

In our aircraft development example, the StRS will contain ACME Air's requirements in the language of the stakeholders. For example, among the many requirements, there will be statements such as the following:

- *The aircraft shall operate from Class X airfields.*
- *The aircraft shall provide class-leading comfort for passengers.*
- *The aircraft shall be turned around to its next flight within 30 minutes.*

Note that the StRS requirement statements are similar in construction and language to those that will follow in the SyRS, but they do not have to be as precise, and the verification requirements can be less stringent. That is, for example, class-leading comfort in our aircraft example is not directly verifiable, but it is sufficiently bounded for the next level of design (conducted by the system designers) to understand what quantitative statements must be made that, when implemented, will lead to "class-leading" comfort.

The StRS also refers to the whole aircraft capability system—that is, the aircraft, air crew, ground crew, simulators, support systems, and training systems. As illustrated in Figure 2.5, each element of the capability will most likely be procured separately and even at a different time—so, while the StRS covers the capability system, it may well lead to several SyRSs, one for each input to the capability.

The potential audience for the StRS includes everyone who has an interest in the delivered system, from the eventual users, operators, and support personnel to the systems engineers and designers responsible for designing, developing, and manufacturing the system as well as the test personnel responsible for confirming that the system is fit for its intended purpose. Each audience set provides a different view of the requirements.

While there may be many stakeholders who have an interest in the development of the StRS, the population of the StRS is generally best conducted by a small team of experienced stakeholder representatives. It is important that all stakeholders contribute to the preparation of the StRS, however, and elicitation/generation techniques such as structured workshops, surveys, interviews, and prototyping are necessary to ensure that all relevant information is considered.

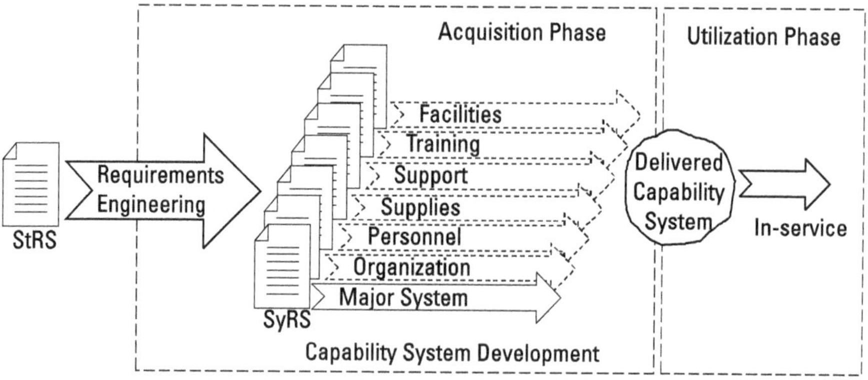

Figure 2.5 An example in which one SyRS is developed for each of the elements of a capability system as described in the StRS.

2.3.2.2 C2.3 Finalize SNR

Finally, the SNR is endorsed by stakeholders and by business management, most likely at a major review. Endorsement of the SNR is a critical step in developing the system in order to provide confidence that all the fundamental stakeholder requirements have been captured. For business analysts and requirements engineers, the SNR is the basis of further development of the requirements to populate the SyRS that forms the centerpiece of the FBL of the system. From a project management perspective, the SNR assists in bounding the preliminary scope of the project from which early cost and schedule estimates can be developed.

2.3.2.3 Traceability

During the subsequent synthesis and analysis, each requirement in the StRS is traceable to at least one requirement in the SyRS. This allows customers to be confident that all StRS requirements have been addressed and gives them insight into the satisfaction of the StRS requirements. This is the start of a cornerstone systems engineering concept called traceability. As discussed in Chapter 1, traceability from the StRS to SyRS is called forward traceability, and backward traceability is from the SyRS back to the StRS, which ensures that all requirements contained in the SyRS are traceable back to at least one StRS requirement. This backward traceability ensures that additional requirements (not formally endorsed by the acquirer) have not crept into the SyRS during the analysis and synthesis process during conceptual design.

Requirements creep must be guarded against as additional, unendorsed requirements ultimately compete with endorsed requirements for valuable design space and may end up reducing the performance of the system against the endorsed requirements. Unendorsed requirements may also add significantly to the cost and schedule associated with the system development, without adding to the system's ability to solve the original stakeholder need.

Managing requirements and any changes to any requirement sets are essential aspects of the project management knowledge area of scope management.

2.4 C3. Define System Requirements

Figure 2.6 illustrates the next step in conceptual design, which is to define the system-level requirements of the system. The system-level requirements form the basis of the functional design of the system—the main aim of conceptual design.

The aim of defining system requirements is to determine what the system must do in order to meet the business and stakeholder needs and requirements—that is, to describe the requirements at the system level and be able

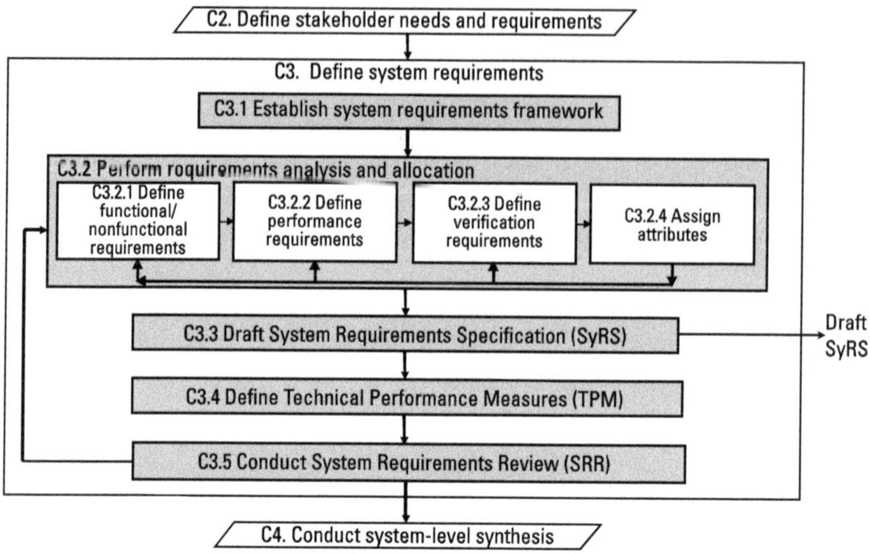

Figure 2.6 A summary of C3.

to relate the functional design back to the BNR/SNR, particularly the StRS. The system requirements definition therefore uses elaboration (decomposition/derivation) to bridge from stakeholder requirements to system requirements.

During system requirements definition it is important to concentrate on what is required rather than how to do it. For example, a system-level performance requirement of the aircraft system might be a specified cruising speed. The physical subsystems that would be needed to perform this function (such as engines, airframes, and flight controls) may not be considered at this stage. The latter stages of the systems engineering processes focus on the physical subsystems (and lower) and how system-level requirements will be met best through a disciplined design approach.

Specifying how to meet the requirements at this embryonic stage may result in ill-informed design decisions leading to suboptimal designs. Because the acquirer is most likely to be performing the conceptual design tasks, specifying at this stage how requirements are to be met will also have the undesirable effect of causing the acquirer to take on some responsibility and risk for achieving system performance; such risk should be taken by the developer.

2.4.1 C3.1 Establish System Requirements Framework

The first step in the systems requirement analysis is to develop a requirements framework (Figure 2.6, activity C3.1) around which the functional design of the system is to be based. Here we call the requirements framework the RBS. The words are deliberately chosen to differentiate this structure from the well-known project management artifact WBS, which includes the system breakdown structure (SBS). The two are not the same. The RBS is grouped by function, while the WBS is structured by physical project elements and contains other project-related information.

The RBS for a system is the structure upon which the many system-level requirements are developed. The RBS framework provides a reference and a guide to those performing the analysis work during the development of the system FBL. In fact, the framework, when completely populated, can form the structure of the SyRS.

Example 2.10: Sample RBS for the Domestic Security Alarm Example

Figure 2.7 depicts a representative requirements framework for our domestic security system example showing the detail completed at the BNR/SNR level.

There are a number of advantages associated with the establishment of a suitable framework such as the RBS early in the requirements analysis activity. First, the framework will act as a reference source during requirements analysis to ensure that all aspects of the system requirements are addressed and that important areas are not omitted. The framework, which is ideally suited to the top-down approach, allows systems engineers to be confident at any level that the needs/requirements at that level are complete, unambiguous, and nonredundant. The framework allows multiple people to work on the analysis simultaneously as it enables the effective allocation of responsibility for sections of requirements to individuals. The framework assists in avoiding duplication of requirements in different sections, which is undesirable as it raises the probability of conflict between requirements and often leads to ambiguity and

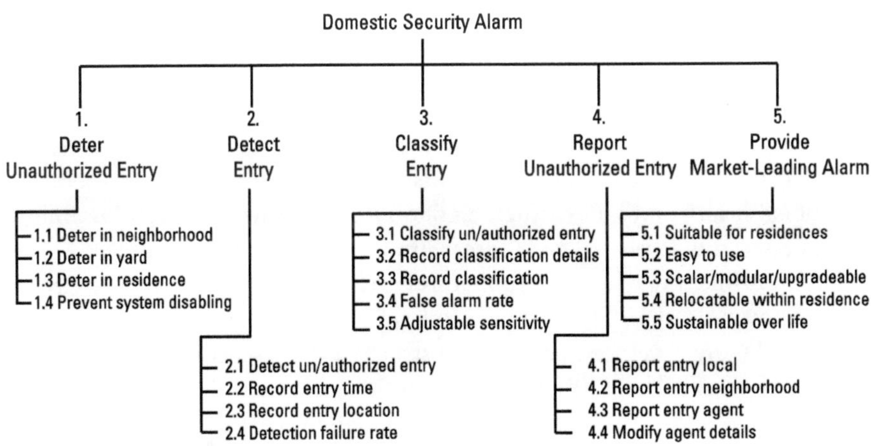

Figure 2.7 Sample RBS for our domestic security alarm example.

confusion. In the worst case, in fact, requirements may end up being instantiated in a duplicated manner when inadvertently allocated to different system elements during preliminary design.

2.4.2 C3.2 Perform Requirements Analysis and Allocation

This activity involves the elaboration of major requirements for the system (with associated performance, verification, and rationale) including necessary functions such as operations, maintenance, and support. When we use the word system from this point forward, we are using the systems paradigm, because, as noted earlier, there will be more than one SyRS produced for each StRS for the capability system (i.e., one for the aircraft, one for the personnel, and one for training).

2.4.2.1 C3.2.1 Define Functional/Nonfunctional Requirements

Defining functional requirements (Figure 2.6, step C3.2.1) is an important part of the overall system requirements analysis that, as the name suggests, identifies the many different functions that the system needs to perform to meet the endorsed mission, goals, and objectives contained in the BNR/SNR (integrated set of needs). The analysis identifies all functions, including maintenance functions that are necessary to maintain or return the system to operational use. At this stage in the development, the concentration is still on what needs to be achieved, not how it should be done.

Operational needs and requirements are normally given the highest priority during the early stages of a project because operational aspects tend to be the most tangible and most easily understood. The life cycle approach also acknowledges that design decisions made early in the acquisition phase can have a large impact on the maintenance and support of the system. To that end, maintenance and support requirements need to be determined and specified during conceptual design so that the design of the system ensures efficient and effective support during the utilization phase.

The consideration of human factors also produces a number of additional prerequisites. Requirements for the system are obtained from an analysis of physical space limitations, climatic limits, eye movement, operator reach, occupational health and safety issues, ergonomics, cognitive limits, and usability. The assessed skill levels of known users also help to constrain the design.

A possible way to analyze functions is via FFBDs [21]. See Figure 2.8 for a simple example of an FFBD shown to level 3.

FFBDs are developed primarily to structure the system requirements into functional terms. The FFBDs identify the major system-level (or top-level) functions that need to be performed by the system in accomplishing its mission. Each top-level block is expanded as necessary to ensure that the functionality

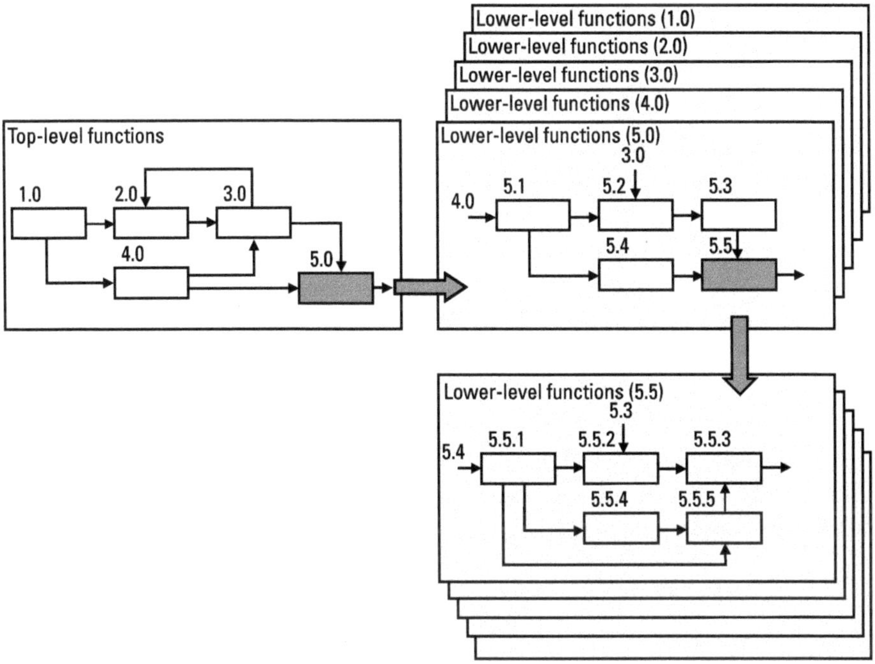

Figure 2.8 Example FFBD (to level 3).

of the system is adequately defined. This may require the functional analysis to break the system down further than the major system-level functions even at this early stage. The requirements should be decomposed as far down as necessary to achieve this aim. FFBDs help designers to evaluate each function adequately and assist in the definitions of the requirements in terms of performance, inputs, outputs, controls, and constraints, which is the next stage in conceptual design.

Once the FFBDs have been completed to the required level, all major system functions and their interrelationships should have been identified. Functional allocation may occur where the functions are grouped together in logical groups and assigned to one of the headings in the RBS. This helps when system-level performance requirements are considered but also ensures that the SyRS that results from the population of the RBS is logical and readable and avoids duplication and potential conflict. This forms a sound framework within which preliminary design can be conducted.

Note that the FFBDs shown in Figure 2.8 make use of a numbering system. The system-level functions have been assigned numbers, and those numbers have been continued through the lower-level functions. In this way, even the lowest level functions and requirements can be traced back to the original system-level function. Traceability has already been introduced as a cornerstone in systems engineering, and it becomes increasingly important as the design matures and becomes more complex.

2.4.2.2 C3.2.2 Define Performance Requirements

Once the functions have been identified and grouped according to the agreed-upon RBS, performance-related parameters must be considered (Figure 2.6, step C3.2.2). Having decided what the system must do, the designer must now determine how well the system is to perform each of those functions. A good practice is to ensure that each time a functional statement is articulated, one or more corresponding performance statements are made.

Most of the operational functions will have obvious performance parameters associated with them such as speed, accuracy, endurance, and acceleration. Support and other functions also require parametric definition to define completely the requirement. Examples of these parametric requirements include environmental considerations, quality factors, support requirements, and utilization requirements.

2.4.2.3 C3.2.3 Define Verification Requirements

The definition of functional and performance requirements is not complete until verification requirements (Figure 2.6, step C3.2.3) are also included. It is good practice to ensure that each time a requirement is articulated, a corresponding verification statement is made. It is often difficult to write verification requirements for system-level functions, but the discipline of doing so is important since there is little point in stating a requirement for a function without consideration of how the function is to be tested. An additional benefit of adhering to the discipline at this stage is that test plans and procedures become much easier to write because the tests required at any level can be considered within a framework provided by an aggregation of the verification requirements at that level.

2.4.2.4 C3.2.4 Assign Attributes

An attribute is additional information associated with a need or requirement that is used to aid in its definition and management. Attributes also aid in achieving many of the characteristics of needs and requirements defined in the INCOSE *Guide to Writing Requirements* [22]. Need statements and requirement statements are supported by associated attributes (Figure 2.6, step C3.2.4) that aid in the definition and management of needs and requirements and of sets of need and requirements [23]. The INCOSE NRM defines 46 attributes (of which 12 are considered mandatory) that can be organized within four broad categories [24], described as follows:

- *Attributes to help define the requirement and its intent:* Examples include rationale, system of interest primary verification method, system of interest verification approach, parent requirement, source, condition of use, and states and modes.
- *Attributes associated with the system of interest verification:* Examples include system of interest verification level, system of interest verification phase, system of interest verification results, and system of interest verification status.
- *Attributes to help maintain the requirements:* Examples include unique identifier; unique name; originator/author; date requirement entered; owner; stakeholders; change board; change status; version number; approval date; date of last change; stability; responsible person; requirements verification status; requirement validation status; status (of requirement); status (of implementation); trace to interface definition; trace to peer requirements; priority; criticality; risk; key driving requirement; additional comments; type/category (including functional/performance; interactions with external systems, input, output, external interfaces, environmental, facility, ergonomic, compatibility with existing systems, logistics, users, training, installation, transportation, storage; the "-ilities" (quality), reliability, availability, maintainability, accessibility, safety, security, transportability, quality provisions, growth capacity; physical characteristics; standards and regulations—policy and regulatory, constraint, business rule, business requirement).
- *Attributes to show applicability and allow reuse:* Examples include applicability, region, country, state/province, application, market segment, business unit, and business line.

For example, it is particularly useful to record the rationale behind each requirement, which will further assist with the removal of ambiguity but more importantly help in situations where more than one person is involved in the requirements analysis over the course of the project. The rationale explains why each requirement is necessary and the logic behind performance imperatives assigned to the requirement. A requirement should not be accepted unless the rationale is well understood. This rationale remains attached to a requirement throughout the acquisition phase, providing guidance to lower level designers as to why and how the requirement has been decomposed/derived and why the performance and verification requirements were selected.

Example 2.11: Utility of Rationale

> *An example commonly used in systems engineering circles involves the design of a back windscreen of a passenger car, which had a user requirement to be able to withstand the air pressure associated with the car moving at 60 mph. During the design, it was established that the car's maximum speed in reverse gear would only be 10 mph, so designers reduced the requirement for the rear windshield to withstand air pressure from 60 mph to 10 mph. Such a decision seemed to make sense, particularly since there was likely to be a subsequent cost saving in the back windscreen component. However, when the new vehicles were reversed onto a car transporter for delivery from the factory to retail outlets, the rationale for the original user requirement became readily apparent when the rear windscreens were subjected to 60 mph on the transporter.*

2.4.3 Notes on Analysis and Allocation

Once functional, performance, and verification requirements have been defined at an upper level, designers begin to analyze and allocate those functions. The aim of requirements analysis is to ensure that the elements of a system-level functional architecture are defined at sufficient depth to allow for synthesis of solutions in the following stage. Recall that the process of requirements analysis is the bridge between stakeholder requirements in the StRS to the system requirements in the SyRS.

The requirements articulated by the stakeholders are generally at a high level and often tend to be more qualitative than quantitative. In fact, the major requirements collected in the first pass will look very similar to the key statements of the StRS. For example, our new aircraft is to operate from any Class X airfield. While stakeholders will articulate some additional requirements during requirements elicitation/generation, these broader functions need to be analyzed to identify the lower level requirements necessary to achieve the parent requirement. These lower level requirements are often called decomposed or derived requirements, because they are not stated directly by a stakeholder but are an artifact of the requirements-analysis process. For example, if the parent function is to be able to operate from any Class X airfield, designers would decompose/derive requirements related to such factors as minimum runway lengths, runway surfaces, and aircraft weight-related restrictions.

The RBS is a very useful tool to support the design and to communicate with stakeholders. However, because it simply states the function required of the system, it is sometimes difficult to determine how those functions relate to each other. The flow of functions is ideally illustrated by FFBDs, which are also very useful tools at this stage, both as analysis tools and as additional mechanisms for communication with stakeholders.

In performing the system requirements analysis, the following tasks are performed:

- Determining the major functions required of the new system (including maintenance, support, and operational);
- Assigning the necessary performance-related parameters or requirements to each of the functions as applicable;
- Articulating any other system-level requirements, such as those in the requirement categories shown in the example RBS in Figure 2.7;
- Determining the technical performance measures (TPMs) to be applied during the systems engineering effort.

Many tools exist to assist with the process of requirements analysis. For example, IEEE-STD-1220 [25] is a systems engineering standard that suggests a procedure for performing requirements analysis. Chapter 6 briefly describes IEEE-STD-1220 and other standards.

Example 2.12: Moving Toward System-Level Requirements in the SyRS

Three possible StRS statements relating to our aircraft example are shown briefly as follows, along with examples of decomposed/derived system-level statements developing the lower level of requirements moving toward the system-level requirements in the SyRS:

1. *Operate from Class X Airfield*
   - 1.1 *Operate with a specified maximum take-off distance*
   - 1.2 *Operate from specified runway surface*
   - 1.3 *Operate with specified maximum allowable aircraft mass*
   - 1.4 *Operate with specified essential navigation aids*
   - 1.5 *Operate with specified essential automatic landing systems*
   - 1.6 *Operate with specified essential communications systems*
2. *Provide class-leading comfort to passengers*
   - 2.1 *Provide class-leading seating*
   - 2.2 *Provide class-leading entertainment systems*
   - 2.3 *Provide class-leading bathroom facilities*
   - 2.4 *Provide class-leading catering services*
3. *Turn the aircraft around in less than 30 minutes*

3.1 Load and unload passengers, cargo, and catering in less than 30 minutes

3.2 Conduct refueling operations in less than 30 minutes

3.3 Conduct operational maintainability in less than 30 minutes

Naturally, there are many requirements needed to define fully the logical design of the aircraft system. Lacking space here to include them all, we list only those sufficient to illustrate the concepts associated with conceptual design and preliminary design. Note that the numbering system has been provided to support traceability throughout the process. For example, the statements associated with 3.3, xonduct refueling operations in less than 30 minutes, relates directly to StRS statement 3.0, turn the aircraft around in less than 30 minutes.

Note that the functions in Example 2.12 are offered to illustrate the decomposition and derivation processes. The functions at the level they are shown are not low enough to be adequate system requirements (i.e., requirements that refer specifically to the system of interest and nothing else). For example, as noted about the FFBD in Figure 2.9, the function 3.3 "conduct refueling" will require a number of actors and other equipment that will be outside the system boundary. The decomposition/derivation process must continue until the

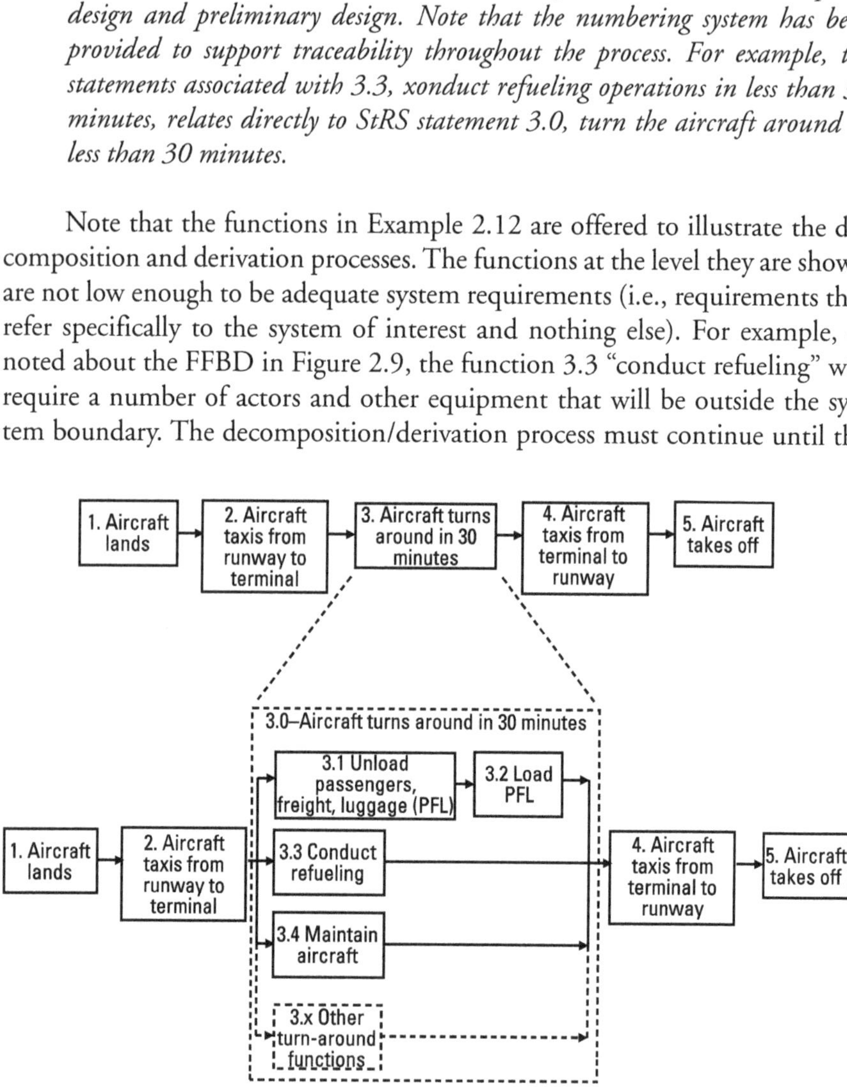

Figure 2.9 An FFBD describing the decomposed/derived functions involved in turning the aircraft around.

system requirements in the SyRS refer only to the aircraft system, not any entity outside the system.

Example 2.13: Use of FFBDs During Conceptual Design

> *During conceptual design, designers use FFBDs to identify system-level functions and to assist in requirements analysis. An FFBD incorporating the turnaround operation of the aircraft may take the form shown in Figure 2.9. "Turnaround" involves processing the passengers, luggage, freight, and so on between flights and preparing the aircraft for its next flight. Note that many such FFBDs are required to cover the other functions required of the aircraft system.*
>
> *The StRS requirement 3.0 to turn the aircraft around results in a number of system-level requirements, including 3.3 "conduct refueling." We can see that 3.1, "conduct refueling," must take no more than 30 minutes to complete. This is an example of a functional requirement (refueling) and an associated performance parameter (less than or equal to 30 minutes).*

Note that, in this example, there are small differences in the terms used in the list of requirement statements (that would be in the RBS) and the functions detailed in the above FFBD. These two tools have different views and are often used simultaneously. We have left the two artifacts in their initial states to illustrate that the two views must be reconciled. For example, the RBS requirement statements refer to passengers, cargo, and catering, whereas the FFBD refers to passengers, freight and luggage. These are similar but have additional information in each, so the designers must reconcile them in order to identify all the functions required (as well as to ensure that, in the end, all functions are referred to in only one way).

In Examples 2.12 and 2.13 we are considering two of the three aspects of allocation introduced earlier. First, requirements are allocated to the appropriate group of the RBS. Second, higher level requirements flow down to lower level requirements. For example, the aircraft must be able to be turned around within 30 minutes. Figure 2.9 shows that this requirement consists of the lower level functions of unload passengers, conduct refueling, and maintain aircraft, which can all be conducted in parallel within the timeframe. We can therefore allocate the performance requirement to lower functions and can allocate the performance measure so that we state that unloading passengers, refueling, and maintenance must each be completed within 30 minutes.

Note that, even at the highest level, the FFBD in Figure 2.9 contains requirements that may not have been directly articulated by the business and stakeholders. For example, the functional flow for this series of activities begins with landing the aircraft, a requirement that is implied in the need to turn the aircraft around in 30 minutes but may not have been stated explicitly. The designers are responsible for ensuring that these implied requirements are captured early in the process and do not lie dormant until identified during user evaluations later in the process.

As a corollary, at this stage in our considerations, the FFBD also still contains functions that will involve users—that is they are not specifically functions of the system. Those functions cannot be carried forward into the SyRS but must be captured separately. As outlined in Chapter 1, for the majority of this text we focus on the acquisition of the major equipment element (often called the material system or the combination of mission system and support system) of the capability system. However, we must always keep in mind, as illustrated in Figure 2.5, that all the capability elements need to be brought back together prior to introduction into service in order to field an operational capability.

2.4.4 C3.3 Draft SyRS

The complete and populated RBS forms the basis for the SyRS, which, when approved, becomes the system's FBL. At this stage, however, the SyRS is only in draft form (Figure 2.6, C3.3), because the requirements collected and analyzed so far have yet to be synthesized into an architectural solution. There will therefore be some degree of latitude allowed in certain requirements, and labels such as mandatory, important, and desirable leave some flexibility for subsequent design.

The SyRS may take any one of many forms [26]; the most appropriate form depends on the application. The SyRS is perhaps the most important of all systems engineering artifacts, because it becomes the source of reference for all the subordinate specifications that are produced at later stages in the design process. A solid SyRS provides an effective foundation for the remainder of the design and development. Errors or omissions in the SyRS flow into the remaining design effort. The later these errors are discovered, the more expensive and time-consuming the rectification becomes.

The endorsed stakeholder requirements in the StRS are expanded in the SyRS to be a set of system-level requirements that, when developed, will result in a system that meets the StRS requirements. However, the SyRS is much more detailed and has transformed the StRS statements in the stakeholders' language into more formal requirements that relate solely to the system. That is, the SyRS, therefore, only includes statements that relate to the system of interest that is the subject of that SyRS—the aircraft, the facility, the simulator, and so on.

Example 2.14: StRS Versus SyRS Language

The language utilized in the StRS is intended for easy reading for a larger audience than the specification language, which tends to be more precise. For example, the StRS requirement statement, "The Aircraft shall operate from Class X airfields" will spawn, among many other requirements, the more detailed SyRS statement, "The Aircraft shall exert less than Y kN/m² of pressure on the tarmac when the aircraft is stationary and at maximum all-up mass."

Note that the aircraft operations may be defined in terms of modes and states. In that case, there may be a Parked_State in which the aircraft is fully loaded and stationary, so that the requirement might then say, "When in the Parked_State, the Aircraft shall exert less than Y kN/m² of pressure on the tarmac when the aircraft is stationary and at maximum all-up mass."

2.4.5 C3.4 Define TPMs

The parametric requirements determined during the preceding stages assist in developing selected quantitative criteria (Figure 2.6, activity C3.4) called TPMs that the system must meet. TPMs are identified early in the system development effort and are continually monitored and tracked throughout the system development as a means of managing the risks associated with system development. IEEE-STD-1220 [27] states that appropriately selected TPMs can be used to assess conformance to requirements, assess conformance to levels of technical risk, trigger development of recovery plans for identified deficiencies, and examine marginal cost benefits of performance in excess of requirements.

The first step in establishing TPMs is to identify the quantitative parameters that require tracking throughout the project. Once the parameters have been identified, they should be prioritized in terms of their importance as viewed by the acquirer. A second list of TPMs may be established, prioritized, and maintained by the supplier. As is often the case with subjective issues such as risk, a second perspective is always valuable. Both lists will imply priority for potential suppliers of the system during its design and development. This level of priority will assist the designer in determining the design emphasis to be placed on achieving each of the TPMs and the potential for trading off the performance of one TPM against the performance of another.

Throughout the design process, a list of TPMs and associated metrics should be maintained along with the priority, the benchmark objective, and the current level of achievement and projected/estimated performance. It is normal for some TPMs to move down in the priority order or drop from the list entirely as the design proceeds, because their ultimate achievement is no longer uncertain. It is also normal for activities that were not considered risky during the early stages of design to be added or moved up in priority due to previously unforeseen complications. This assists both the acquirer and the supplier in the identification and mitigation of technical risks during the acquisition phase.

Bands of acceptable variation from the expected value will also be determined and agreed upon so that unnecessary risk management actions are not instigated at the first sign of a very minor variation. The variation bands will typically be large in the early stages of the design but will become increasingly narrow as the design matures.

Example 2.15 may clarify the role of TPMs in the acquisition phase of the systems engineering effort and emphasize the uses of TPMs as described by IEEE-STD-1220.

Example 2.15: TPMs for Our Aircraft Example

Our aircraft system must be capable of a minimum rate of climb in order to comply with regulatory requirements governing departures from airfields. The designers of the aircraft system may identify rate of climb as a major design challenge and therefore a candidate for tracking using the TPM process. Designers can estimate the likely rate of climb of an aircraft system of this type, but their estimate will not be confirmed until the design of the aircraft matures.

In the early stages of a system life cycle, it is reasonable to expect that the aircraft design will exist on paper and in computer models, allowing an initial estimate of rate of climb to be made. The design will continue as the subsystems are designed and incorporated into the overall airframe design. Each subsystem will add mass to the aircraft and may change the aerodynamic performance of the aircraft (for example, external antenna systems). To that end, the subsystem design will start to alter the rate of climb that the aircraft is likely to achieve, thereby refining the initial estimate. During the latter stages of design, a prototype aircraft may be produced and test-flown. Among other things, this test flying will confirm the validity of the earlier estimates of rate of climb and establish actual rates of climb achievable under different circumstances.

To monitor the risks associated with this critical TPM, the aircraft rate of climb could be estimated, investigated, and reported upon at discrete stages throughout the acquisition phase. Design reviews are a convenient forum for such activities. The progress of rate of climb toward the required rate of climb is of interest to both the designers of the system and the customers of the system. Failure of the aircraft system to meet this requirement will prevent the aircraft from meeting all of its original (and endorsed) needs, goals, and objectives (in this case, the requirement to operate from Class X airfields).

The graphs in Figure 2.10 and Figure 2.11 show two possible profiles of rate of climb when 50% of the system design is complete. Although both profiles show the same rate of climb estimate at the 50% mark, the likely outcome from each scenario is very different.

In Figure 2.10, the rate of climb is decreasing as additional subsystems are designed and added to the aircraft. The decrease is staying well within the tolerance band and is tending toward the minimum acceptable rate of climb

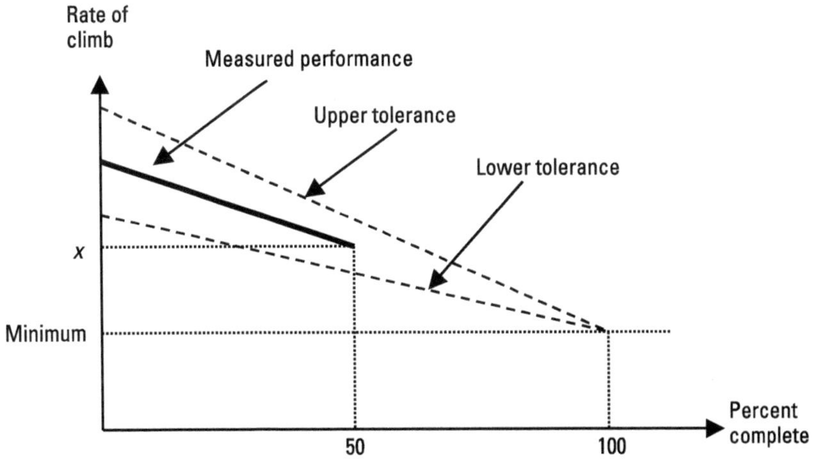

Figure 2.10 Rate-of-climb TPM as a function of percentage of work complete—Example 1.

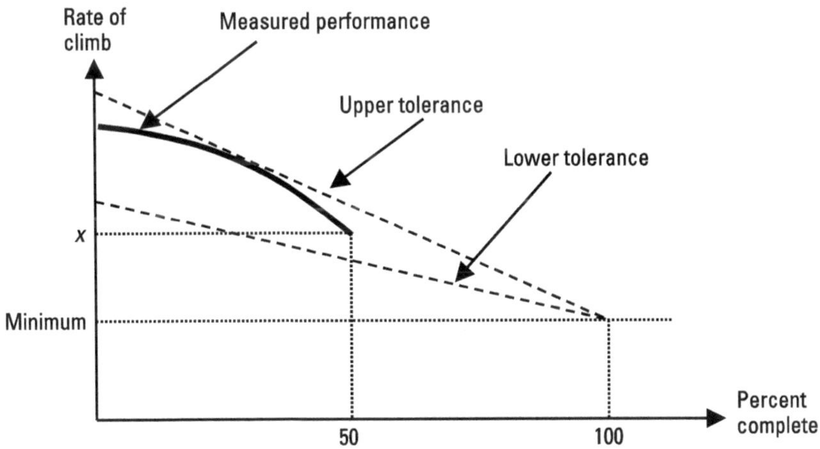

Figure 2.11 Rate-of-climb TPM as a function of percentage of work complete—Example 2.

in a linear (predictable) fashion. At the 50% mark, the aircraft system has an estimated rate of climb of x feet per second. In this scenario, the rate-of-climb TPM would be presented at the relevant design review, and it is likely that the reviewers would not see any real problems with rate-of-climb TPM.

In Figure 2.11, the rate of climb estimate at the 50% complete mark is also x feet per second. If the instantaneous estimate was the only indicator of rate of climb, it would seem reasonable enough. However, when the trend is observed, the rate of climb estimate brushes the upper tolerance band and then starts to dip alarmingly toward the lower tolerance band. Presented with this scenario, reviewers should note the trend toward the lower tolerance band and raise some concerns. In all likelihood, there may be a reasonable explanation for the trend. For example, the heaviest subsystems may have been designed during the first 50% of the project, and the remainder of the subsystems are not expected to detract too much further from rate of climb. However, if there is no reasonable explanation, more detailed investigation is required to support risk management.

TPM tracking should occur throughout the acquisition phase, and the parameters should be assessed periodically during technical reviews. TPM management (including identification, prioritization, and setting of values) is therefore an important part of conceptual design.

2.4.6 C3.5 Conduct System Requirements Review

System requirements reviews (SRRs) may be conducted periodically throughout conceptual design to verify and approve sets of system-level requirements. The number of SRRs depends on the size and complexity of the system. Large and complex systems may require multiple SRRs during conceptual design, whereas simple systems may not require any. It is recommended that at least one SRR is held, however, even in simple projects, so that the stakeholders have some formal involvement in the transition from the StRS to the SyRS. In some organizational development methodologies, there is only one SRR held before tendering.

The aim of SRRs is to monitor and approve progressively the system-level requirements that are developed on the way to the FBL. Progressive reviews allow the requirements analysis effort to continue with confidence in the preceding decisions. SRRs may or may not be considered formal reviews.

Requirements reviewed at this stage are likely to contain margins of acceptable performance and may be categorized according to priority. A possible categorization may see requirements listed as mandatory, important, or desirable where mandatory requirements are absolutely essential, desirable requirements are nice to have, and important requirements are somewhere in between. Each SRR will review the margins and category of each requirement to ensure that each is being afforded appropriate weighting. Other artifacts such as manufacturing plans, design schedules, and personnel requirements plans are also reviewed at the SRR.

2.4.7 Management Considerations

Although the system-level requirements are captured in the draft SyRS, there are additional project management and systems engineering considerations that need to be captured in terms of how the system development is to be conducted and managed. The following nonexhaustive list provides illustrations:

- Responsibilities for project management, systems engineering, logistics management, operation, and support should be determined and clearly articulated so that all stakeholders are aware of their responsibilities at all stages of the system life cycle.
- Project management considerations will include principal responsibilities, stakeholder engagement, schedule issues, budget issues, risk management, project management milestones, and payment schedules.
- Systems engineering aspects will include expectations for test and evaluation, configuration management, artifact deliverables, technical reviews and audits, and adherence to standards.

These requirements need to be identified and planned, because they will dictate how the design and development effort will be organized and managed.

The preceding examples (and many more) are necessary steps to reduce the technical risk of the development effort, but they require work and therefore must be accounted for by the project in the WBS.

Some management activities will be covered by work conducted in accordance with internal acquirer processes and procedures. When development of the system is to be outsourced, however, the management aspects must be contained in a contractual artifact such as an SOW. Broadly then, in a customer-contractor relationship, the central artifacts in a contract will be the SyRS, which articulates what is required of the system to be delivered by the contract, and the SOW, which describes the work to be conducted by the contractor during the contract.

2.5 C4. Conduct System-Level Synthesis

System design has now progressed to the stage where some of the system-level design decisions can be made based on the draft system requirements. Synthesis at the conceptual design level establishes a system configuration that is representative of the final system form at the subsystem level. At this point, the configuration is not assumed to be final, since the design may be adjusted later in the design process as the design becomes physical.

In system-level synthesis, potential system solutions are developed (Figure 2.12, activity C4.1). Information is collected (Figure 2.12, activity C4.2) regarding how the potential solutions meet the draft SyRS. The potential solutions are then evaluated (Figure 2.12, activity C4.3) by the acquirer, which requires the development of suitable evaluation criteria as well as an evaluation framework. Any discrepancies in the systems engineering products are noted) and fed back to previous activities. The preferred solution is chosen (Figure

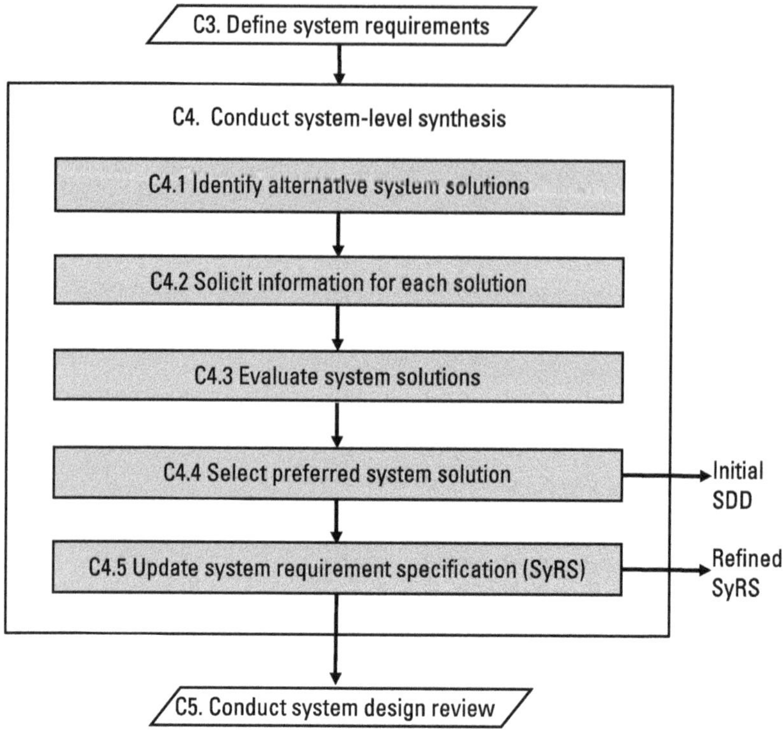

Figure 2.12 A summary of C4. Conduct system-level synthesis.

2.12, activity C4.4), and the system architecture is generally captured in a system design description (SDD) [28] or equivalent, which is used to show compliance against the requirements in the draft SyRS.

Although system-level synthesis can be conducted internally by the acquirer, it often takes the form of a request-for-tender (RFT) in which tenderers are provided with a draft SyRS and an SOW. Each tender response shows how the solution offered provides a feasible system-level design that meets (complies with) the SyRS and how the work requested in the SOW will be conducted. The acquirer evaluates the tender responses before selecting the preferred system-level solution and inviting the preferred tenderer to negotiations. If negotiations are successful, a contract is signed.

The draft SyRS will contain margins of acceptable performance and priorities providing a sufficiently broad design space to allow a range of solutions to be tendered. As part of tender evaluation (synthesis), the draft SyRS is tested against a range of feasible system solutions and a preferred solution selected. The draft SyRS requirements now need to be refined to remove the ranges of acceptable performance and to reflect more precisely what is to be achieved by the selected system and the minimum acceptable level of performance. New system-level requirements may need to be added to the SyRS, and errors or conflicts in existing requirements need to be clarified. The priority of requirements is often removed so that each requirement statement is essential. As feasible alternatives are presented, the acquirer may also need to revisit the BNR/SNR and seek endorsement from the business/stakeholders for proposed changes to the draft SyRS if any of the endorsed requirements are not feasible within the approved schedule or cost constraints.

So, the final activity in process C4 (Figure 2.12, activity C4.5) refines the SyRS, updates BNR and SNR, and establishes the initial FBL. The main products of synthesis are the refined SyRS and the broad system solution described in the SDD.

The SyRS is perhaps the most important systems engineering artifact because it becomes the source of reference for all the subordinate specifications that are produced during later stages of the design process. A solid SyRS forms an effective foundation for the remainder of the design and development effort. Errors or omissions in the SyRS flow into the remaining design effort. The later these errors are discovered, the more expensive and time-consuming the rectification will be. Additionally, the SOW is perhaps the most important project management artifact as the work described in it will be translated into the work packages in the developer's WBS; again, any errors or omissions will have consequences in terms of the project budget and schedule.

2.6 C5. Conduct SDR

At the end of conceptual design, the SDR provides the following (from a systems engineering perspective):

- Formal confirmation that the logical design meets the business and stakeholder requirements;
- Formal record of design decisions and acceptance;
- Formalized communication of the intended design approach, as recorded in the SDD or equivalent, to the major players in the design effort;
- Approval of the verification and validation plans for the system;
- Approval of the systems engineering management plan (SEMP) and supporting plans.

In addition to review of the systems engineering effort, a number of review activities will most likely be performed at SDR to support project management:

- Confirmation that the system to be procured aligns with the customer's organizational goals;
- Refinement of the SOW and a better understanding of the work required;
- Refinement of the project management plan (PMP);
- Refinement of other early plans, such as production plans;
- Refinement of cost estimates;
- Refinement of the schedule and confirmation that it is consistent with the cost and risk goals for the project;
- Confirmation that all required project resources are available.

Figure 2.13 summarizes the activities involved in the SDR process. As with all reviews, it is essential to establish the entry and exit criteria for the review—that is, it must be clear what must be in place before the review begins and what must have been achieved before the review is considered to be

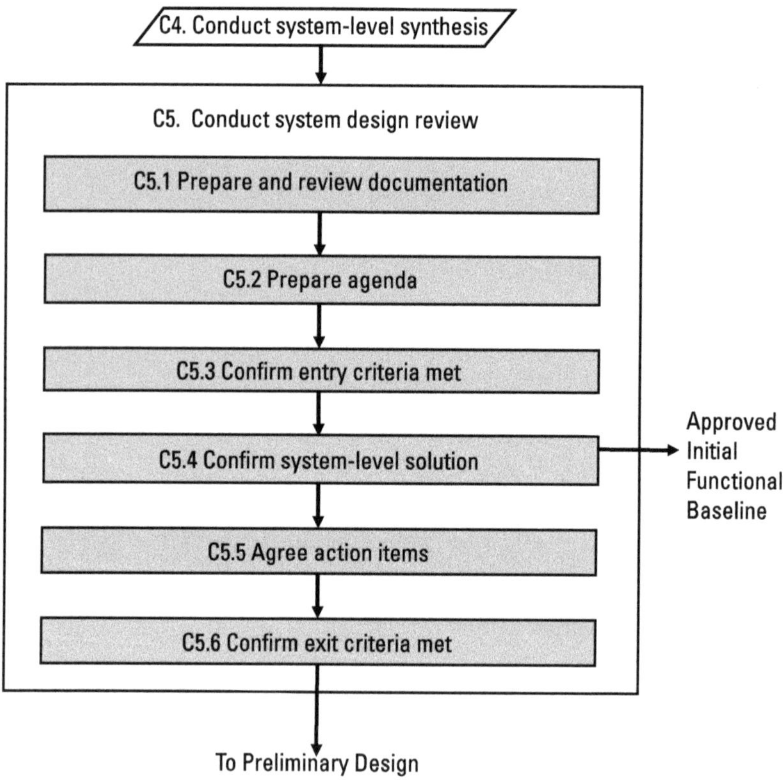

Figure 2.13 A summary of C5. Conduct system design review.

complete. Consequently, the review cannot begin until the entry criteria have been met (Figure 2.13, activity C5.3). Part of that confirmation is to ensure that all of the necessary artifacts have been prepared and reviewed (Figure 2.13, activity C5.1) in advance of the review, and that all parties have agreed to the agenda for the review (Figure 2.13, activity C5.2).

SDR then confirms that the system solution described in the SDD meets the requirements of the refined SyRS (Figure 2.13, activity C5.4). If all is satisfactory, the SyRS is approved and baselined as part of the initial FBL. It is rare that SDR ends with all issues resolved during the review since a number will require some investigation and analysis. Rather than hold up subsequent design, it is common for a number of action items to be agreed on (Figure 2.13, activity C5.5) with regard to outstanding action (provided, of course, that resolution of those issues will not affect the remainder of the design). These outstanding actions are then completed in parallel with the early preliminary design activities and are reviewed for completeness in conjunction with a subsequent review or audit. Before the review can be considered to be complete, the exit criteria must be met (Figure 2.13, activity C5.6).

The boundary between conceptual design and preliminary design is often not clear. From a system design perspective, SDR aims to provide demarcation by approving the SyRS and establishing the initial FBL from which the remainder of the design will follow. For simplicity, we are focused on the SyRS, but we should also note that the initial FBL includes considerably more than the SyRS in terms of systems engineering and project management artifacts. At the very least, there will also be the PMP, the SEMP, the major plans outlined in Chapter 5, and production plans, among others. Consequently, one of the other major aims of SDR is to mark the transition of responsibility from the customer/acquirer/tasking-activity to the contractor/developer/performing-activity.

2.7 Summary

This chapter has outlined the major processes of conceptual design, which is the most important of all activities in the system life cycle, because it starts with a short mission statement and ends with an expansive functional description of the system (in the SyRS) that meets the need. It follows, therefore, since the remaining activities in the acquisition phase are traceable back to the SyRS (particularly in a contracted development), that it is the single most critical artifact in the system life cycle—any errors or omissions in the SyRS will invariably result in errors or omissions in the resultant system when it is fielded. The later these areas are discovered, the more expensive and time-consuming the rectification becomes.

The initial FBL is also critical because it represents the transition from the customer/acquirer/tasking-activity to the contractor/developer/performing-activity—that is, the responsible parties change. It follows that the handover from one to the other, particularly in a formal contracting environment, must be unambiguous and correct if the subsequent work is to be correct.

From a project management perspective, the initial FBL represents the starting point for the formalization of the project scope, which then allows the initial budget and schedule to be developed. Before the scope, budget, and schedule can be finalized, however, the WBS must be developed, which requires a functional-to-physical transformation from the functions in the SyRS and SOW to the contracted WBS. That is the principal purpose of the next activity in the acquisition phase—preliminary design—as discussed in Chapter 3.

Review Questions

1. Define what is meant by a concept, a need, and a requirement.
2. Briefly describe the role and CONOPS and the OpsCon.
3. Briefly describe the business needs and requirements definition process, including the roles of the PLCs and the BRS.
4. Briefly describe the stakeholder needs and requirements definition process, including the roles of the LCs and the StRS.
5. Briefly describe the system requirements definition process, including the roles of the SyRS.

Endnotes

[1] Ryan, M. J., and L. S. Wheatcraft, "On a Cohesive Set of Requirements Engineering Terms," *Systems Engineering,* 2017.

[2] Ryan, M. J., and L. S. Wheatcraft, "On a Cohesive Set of Requirements Engineering Terms," *Systems Engineering,* 2017.

[3] Wheatcraft, L. S., M. J. Ryan, and T. E. Katz, *Needs and Requirements Manual: Needs, Requirements, Verification, Validation Across the Lifecycle,* Hoboken, NJ: John Wiley & Sons, 2024.

[4] International Institute for Business Analysis, *A Guide to the Business Analysis Body of Knowledge® (BABOK® Guide),* Version 3, 2015.

[5] ISO/IEC/IEEE 15288, *Systems and Software Engineering—System Lifecycle Processes,* 2023.

[6] ISO/IEC/IEEE 29148 *Systems and Software Engineering—Lifecycle Processes—Requirements Engineering,* 2018.

[7] ISO/IEC/IEEE 15288, *Systems and Software Engineering—System Lifecycle Processes,* 2023.

[8] ISO/IEC/IEEE 29148 *Systems and Software Engineering—Lifecycle Processes—Requirements Engineering,* 2018.

[9] ANSI/AIAA G-043B-2012, *Guide for the Preparation of Operational Concept Descriptions,* American National Standards Institute, American Institute of Aeronautics and Astronautics (sponsor), 2012.

[10] ISO/IEC/IEEE 29148 *Systems and Software Engineering—Lifecycle Processes—Requirements Engineering,* 2018.

[11] International Institute for Business Analysis, *A Guide to the Business Analysis Body of Knowledge® (BABOK® Guide),* Version 3, 2015.

[12] ISO/IEC/IEEE 29148 *Systems and Software Engineering—Lifecycle Processes—Requirements Engineering,* 2018.

[13] ISO/IEC/IEEE 29148 *Systems and Software Engineering—Lifecycle Processes—Requirements Engineering,* 2018.

[14] International Institute for Business Analysis, *A Guide to the Business Analysis Body of Knowledge® (BABOK® Guide),* Version 3, 2015.

[15] MIL-STD-499B, *Military Standard—Systems Engineering—Draft,* Washington, DC: US Department of Defense, 1994.

[16] ISO/IEC/IEEE 29148 *Systems and Software Engineering—Lifecycle Processes—Requirements Engineering,* 2018.

[17] Wheatcraft, L. S., M. J. Ryan, and T. E. Katz, *Needs and Requirements Manual: Needs, Requirements, Verification, Validation Across the Lifecycle,* Hoboken, NJ: John Wiley & Sons, 2024.

[18] Wheatcraft, L. S., M. J. Ryan, and T. E. Katz, *Needs and Requirements Manual: Needs, Requirements, Verification, Validation Across the Lifecycle,* Hoboken, NJ: John Wiley & Sons, 2024.

[19] Wheatcraft, L. S., M. J. Ryan, and T. E. Katz, *Needs and Requirements Manual: Needs, Requirements, Verification, Validation Across the Lifecycle,* Hoboken, NJ: John Wiley, & Sons, 2024.

[20] Wheatcraft, L. S., M. J. Ryan, and T. E. Katz, *Needs and Requirements Manual: Needs, Requirements, Verification, Validation Across the Lifecycle,* Hoboken, NJ: John Wiley & Sons, 2024.

[21] Blanchard, B., and W. Fabrycky, *Systems Engineering and Analysis,* Upper Saddle River, NJ: Prentice-Hall, 1998. Note that these generic FFBDs can be prepared by any one of the graphical methods including the Integrated Definition (IDEF) modeling method, the Behavioral Diagram Method, and the N2 Charting Method, which are compared and discussed in: Blanchard, B., D. Verna, and E. Peterson, *Maintainability: A Key to Effective Serviceability and Maintenance Management,* New York: John Wiley & Sons, 1995. The DSMC management guide (Defense Systems Management College (DSMC), Systems Engineering Management Guide, Washington, DC: US Government Printing Office, 1990, pp. 6.3–6.5) explains how FFBDs can be used during function identification.

[22] International Council on Systems Engineering (INCOSE), INCOSE-TP-2010-004-04, *Guide to Writing Requirements,* 2024.

[23] Wheatcraft, L. S., M. J. Ryan, and T. E. Katz, *Needs and Requirements Manual: Needs, Requirements, Verification, Validation Across the Lifecycle,* Hoboken, NJ: John Wiley & Sons, 2024.

[24] Wheatcraft L. S., M. J. Ryan, and J. Dick, "On the Use of Attributes to Manage Requirements," *Systems Engineering,* Vol. 19, No. 5, September 2016, pp. 448–458.

[25] IEEE-STD-1220, *IEEE Standard for Application and Management of the Systems Engineering Process,* New York: IEEE Computer Society, 2005.

[26] Sources of information on the content of the SyRS can be found in the following:

Data Item Description (DID), *System/Subsystem Specification (SSS),* DI-IPSC-81431, DD Form 1664, US Department of Defense, April 1989.

ECSS-E-10A, *Space Engineering—Systems Engineering,* Noordwijk, The Netherlands: European Cooperation for Space Standardization, 1996.

MIL-STD-961E, *Department of Defense Standard Practice for Defense and Program-Unique Specifications—Format and Content,* Washington, DC: US Department of Defense, August 1, 2003.

MIL-STD-490A, *Military Standard-Specification Practices,* Washington, DC: US Department of Defense, 1985.

[27] IEEE-STD-1220, *IEEE Standard for Application and Management of the Systems Engineering Process,* New York: IEEE Computer Society, 2005.

[28] Information on the content of the SDD can be found in Data Item Description (DID), *System/Subsystem Design Description (SSDD),* DI-IPSC-81432, DD Form 1664, US Department of Defense, April 1989.

3

Preliminary Design

3.1 Introduction

Preliminary design starts with the initial functional baseline that has been defined during conceptual design and continues by translating system-level requirements into design requirements for the physical subsystems/CIs that will be integrated to form the system. This translation continues the requirements analysis started in conceptual design, followed by the determination of specific requirements for the hardware, software and interfaces making up the system. Trade-off studies are conducted, and the preliminary design effort results in the establishment of an allocated baseline, so named because the system-level requirements are being allocated to the specific subsystems making up the system.

Responsibility for performing preliminary design normally rests with the developer, who accepts responsibility for the system meeting the requirements of the FBL (normally prepared by the acquirer). The acquirer's role now increasingly becomes one of monitoring, reviewing, and supporting the developer's progress. The acquirer normally avoids becoming actively involved in design decisions made during preliminary design (or any subsequent activity, for that matter), since the responsibility for the resultant functionality and performance of the system rests with the developer.

In many cases, the acquirer may opt to add another level of rigor to the process by engaging independent systems engineering consultants to provide independent review of the many and varied engineering documents, artifacts, and products produced by developers during a systems development.

The processes completed during the preliminary design effort include subsystem requirements analysis; subsystem requirements allocation; subsystem synthesis and evaluation; and PDR. Figure 3.1 illustrates these tasks are illustrated in Figure 3.1, which are described in Sections 3.2 to 3.7.

3.2 P1. Conduct Subsystem Requirements Analysis

Requirements analysis has already been described as the process through which requirements are progressively decomposed/derived from the system level to the subsystem level and so on until all functions, parameters, and interfaces have been defined, and the necessary resources have been identified to meet the original system requirements. Subsystem requirements analysis in preliminary design is aimed at continuing the effort commenced during conceptual design to the next level of detail. Figure 3.2 describes the major activities in subsystem requirements analysis.

Because preliminary design will most probably be conducted by a different team than the one that conducted conceptual design (that is, by the developer rather than the acquirer), the subsystem requirements analysis activity begins with a review of the SyRS (Figure 3.2, activity P1.1), and a review of the relevant TPMs contained therein (Figure 3.2, activity P1.2).

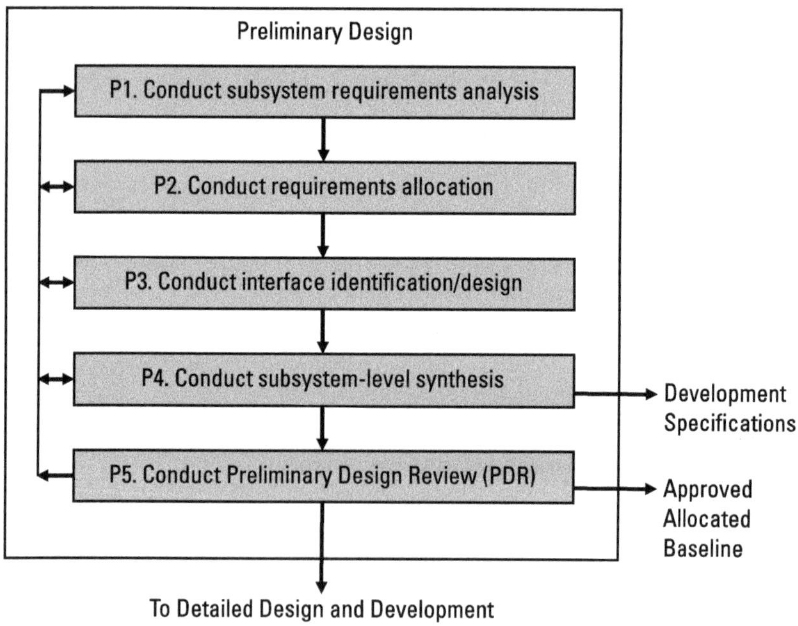

Figure 3.1 Preliminary design processes.

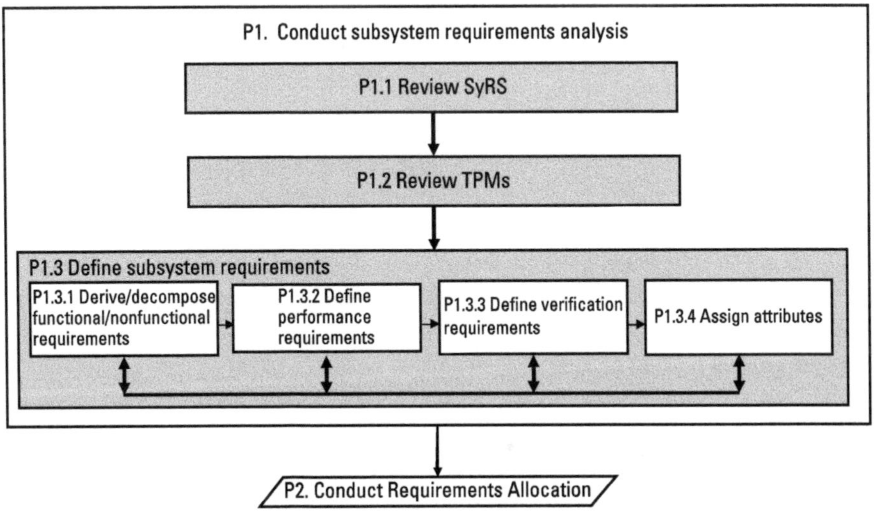

Figure 3.2 Subsystem requirements analysis activities.

The next step is to begin defining requirements at the subsystem/CI level (Figure 3.2, activity P1.3) by decomposing/deriving requirements from the SyRS (Figure 3.2, step P1.3.1) to provide more detailed requirement statements that can be allocated to physical subsystems/CIs. Performance requirements (Figure 3.2, step P1.3.2) and verification requirements (Figure 3.2, step P1.3.3) are also assigned, as are attributes (Figure 3.2, step P1.3.4) for each requirement.

It is not possible to prescribe the level of detail that needs to be decomposed/derived during subsystem requirements analysis. The detail must be sufficient both to define completely the function/performance/constraints associated with the major subsystems comprising the system and to be able to be used either to procure, or to design and produce, the major subsystems and their assemblies and components during subsequent stages of the systems engineering process. The level of detail that is required during preliminary design will therefore be influenced by whether the design is likely to use off-the-shelf items or whether it will require new or modified items to be developed.

FFBDs can be used to support this requirements-analysis process. Each block of the FFBD is eventually considered individually in determining the necessary inputs, the expected outputs, the constraints placed on the function, and finally the mechanisms by which the function could be achieved. Note that this stage is the beginning of the conversion of *what* the system needs to do (contained in the SyRS) into *how* it is going to do it (contained in the SDD and the subsystem development specifications), which is more commonly referred to as the translation of functional design into physical design. Sometimes, as described in Section 3.3, a number of different functions may be performed by the same piece of equipment or piece of software and so on.

Example 3.1: Detailed Aircraft System FFBD

This example illustrates how an FFBD may be used during preliminary design to define system functionality at varying levels of detail.

In Section 2.4.3, we suggested that an FFBD describing the actual turnaround operation of the aircraft may take the form shown in Figure 3.3 (a copy of Figure 2.9). The StRS requirement 3.0 to turn the aircraft around resulted in a number of decomposed/derived system-level requirements, including 3.3 conduct refueling.

Let's investigate this function further as we would during preliminary design. By using the FFBD process to investigate the requirement further in Figure 3.4 we can see that refueling an aircraft involves a number of steps prior to and after the actual refueling process. We can also see that once we have taken the other activities into account, the actual refueling process 3.3.6 (refuel aircraft) must now take place in less than 17 minutes. In the worst case, therefore, we must be able to refuel the aircraft from a completely empty state to maximum fuel capacity in less than or equal to 17 minutes.

The requirement to refuel in less than or equal to 17 minutes has been derived using the FFBD process and is an example of allocation of the upper-level performance requirement of 30 minutes down to 17 minutes for the

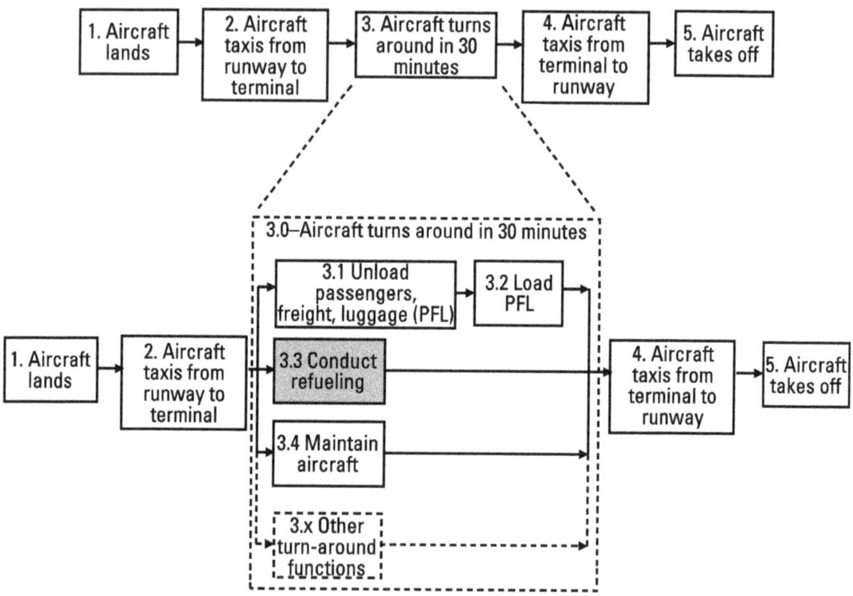

Figure 3.3 An FFBD describing aircraft turnaround.

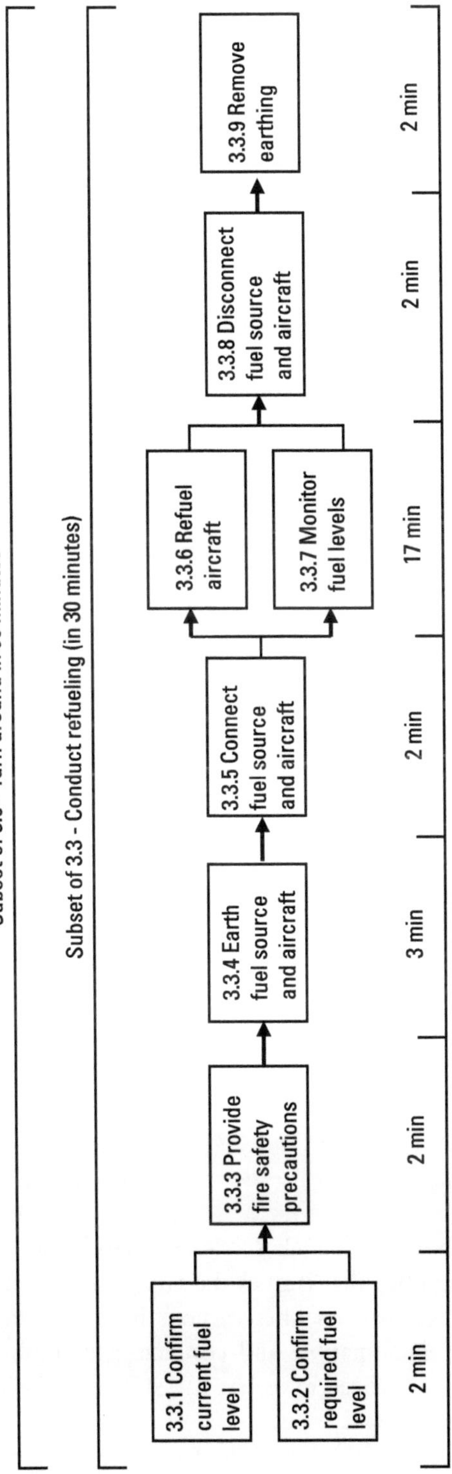

Figure 3.4 An FFBD describing the function 3.3 Conduct refueling.

refueling process. We do not have space here, but in a real project we would continue the FFBD process further and investigate what is involved in the refueling function (3.3.6), and so on.

Note that the numbering of blocks is hierarchical in nature and allows component-level functions to be related through subsystem functions to system-level functions. For example, the numbering system makes it easy to associate function 3.3.3 (provide fire safety precautions) with its parent 3.3 (conduct refueling) and its parent function 3.0 (turn around in 30 minutes).

This example has demonstrated how StRS and SyRS requirements could be investigated further during preliminary design using FFBDs. The requirement will be satisfied by whatever subsystem of the aircraft design is responsible for fuel-related function and performance. This may be called the Fuel Subsystem. Determining what physical subsystem will satisfy each requirement is the start of the translation of functional design into physical design, which is one of the key activities of preliminary design.

Some additional points to note:

- *If a refueling time of 17 minutes is impossible to achieve, some additional time may be gained by investigating the other functions involved in the 3.3 (conduct refueling) function. It may be possible to conduct some functions in parallel, or it may be possible to speed up some functions. Either way, additional time may be provided for the 3.3.6 function. If it is still impossible to perform the task in the allotted time, at least our traceability allows us to trace back up to the StRS requirement to turn the aircraft around in 30 minutes and revisit this imperative (if that is possible).*
- *The FFBD process has also highlighted some interfaces that will need to be considered later in the process. It is worth looking at each of the functions in 3.3 (conduct refueling) and identifying the interfaces. Remember that interfaces include human machine interfaces (HMIs) not just physical interfaces. An HMI will exist, for example, to allow the current fuel level in the aircraft to be determined (3.3.1). A physical interface will exist to allow the aircraft to be correctly grounded (3.3.4) and allow the fuel source to connect to the aircraft (3.3.5).*

It is vitally important that all subsystem requirements analysis is performed from the same FBL, that is from the same SyRS. Although this sounds obvious, different functional baselines versions have been known to exist on projects, resulting in incompatible and conflicting preliminary and detailed designs. This reinforces the need to review the SyRS and the TPMs before embarking on preliminary design.

Once satisfied with the level of decomposition/derivation, each functional block can be considered independently as was the case during system-level requirements analysis. This consideration should focus on the necessary inputs to the function, the expected outputs from it, and any parametric requirements associated with the function. Some thought may also be given to how the function may be performed in reality and what constraints exist in terms of the function's design.

3.3 P2. Conduct Requirements Allocation

As illustrated in Figure 3.5, requirements allocation refers to the process of grouping or combining similar functions and parametric requirements (identified during the requirements analysis) into logical subdivisions. We have done this before during conceptual design, where the groups were based on the RBS headings. During preliminary design, the groups are based around a preliminary physical architecture formulated by the designers, rather than functional RBS headings. The groups of similar functions then assist the designer to determine the design of the major subsystems and ultimately components that are required to make up the system. This represents the translation from functional design to physical design. These subsystems and components are eventually described by either individual CIs or groups of CIs, as discussed later. It is the assignment or allocation of functions to physical CI that is the major focus of requirements allocation.

The first step in requirements allocation during preliminary design is to investigate each of the subsystem-level requirements and the associated

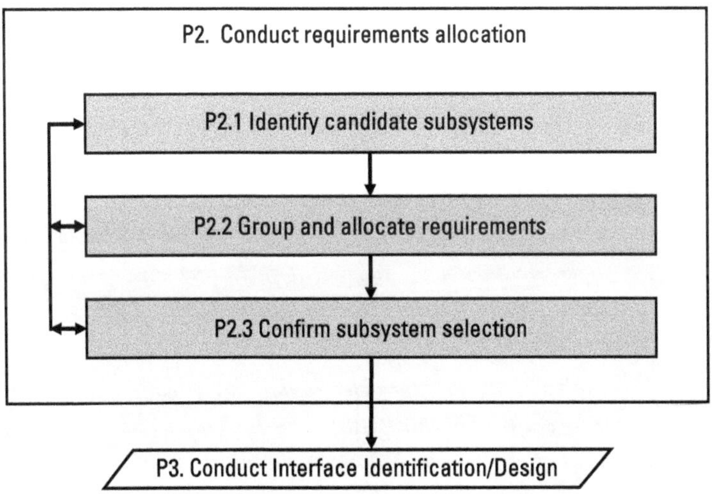

Figure 3.5 The requirements allocation process.

requirements and TPMs flowing from subsystem analysis and group them into logical sets or groups based on the preliminary physical architecture. The logic used relies on experience and expertise in the relevant design domain. In our example, this domain is the aerospace design domain. That is, to perform requirements analysis and allocation during preliminary design in our example, we rely on knowledge of the likely physical composition (subsystems) of the aircraft system. Even though the detailed design is yet to be determined, experienced designers have the elements of a preliminary design in mind.

Example 3.2: Subsystems for Our Aircraft System

Let's now look at the subsystems that comprise any aircraft. (Remember that our aircraft example has been pitched at a sufficiently general level to allow anyone familiar with aircraft to follow the general example.) First, it is worth looking at what the aircraft industry calls the airframe, which consists of a range of subsystems, described as follows:

- *The undercarriage subsystem includes the structures, tires, and brakes necessary to support the aircraft while on the ground.*
- *The wings/fuselage subsystem includes all the components necessary to accommodate the remainder of the subsystems and to form the aerodynamic portion of the aircraft system.*
- *The fuel subsystem includes all the tanks, refueling/defueling receptacles, fuel pumps, and fuel lines necessary to support the engines.*
- *The hydraulic subsystem includes all the hydraulic pumps, lines, and accessories needed to provide hydraulic power to all other aircraft systems such as flight controls and undercarriage.*
- *The flight control subsystem includes all the primary and secondary flight controls necessary for the control of the aircraft system in the air and on the ground.*
- *The engine subsystem consists of the engine itself including all the components that come together to form each individual engine such as turbines, gearboxes, and so on. The engine will also interface to an engine-management subsystem including all the peripheral controls necessary to manage the engine during start-up, operation, and shutdown.*
- *The avionics (or aviation electronics) subsystem assists in the control of the aircraft, performing functions such as electrical generation and distribution, communications, and navigation. Most modern avionics subsystems comprise the following:*

    - *A computer (or collection of computers) that includes the hardware and software necessary to perform all aircraft operational functions;*
    - *Navigation subsystems that include the hardware and software necessary to perform the navigation functions for the aircraft;*
    - *Communications subsystems that include the hardware and software necessary to perform the communications functions for the aircraft;*
    - *Automatic flight-control subsystems that include the hardware and software necessary to perform the flight-control functions such as automatic pilot;*
    - *An electrical generation and distribution subsystem responsible for the generation and distribution of all necessary electrical power to the aircraft systems.*
- *The interior subsystem of the aircraft system will comprise the following:*
    - *The cockpit, which includes all the elements to support the aircrew in their operation of the aircraft;*
    - *The passenger space, which includes all of the elements necessary to support the passengers throughout their flights;*
    - *The cargo space, which includes all of the elements necessary to support the cargo throughout its flights;*
    - *The galley space, which includes all of the elements necessary to support the provision of food, drink, and refreshments to crew and passengers throughout their flights.*

Each of these major subsystems needs to be considered individually during preliminary design. These subsystems will end up consisting of one or more individual design building blocks, known as CIs, depending on decisions made by the system designers. CIs comprise hardware, software, or a combination of both designed to satisfy an allocated group of functions and requirements. As the name suggests, the configuration of each CI is managed as a separate item meaning that different CIs may be at different revision levels as the systems engineering process proceeds.

The selection of physical design items and their designation as CIs is a major part of the overall configuration management process called configuration identification (detailed in Chapter 5). In general, though, items may be identified as CIs because of the following factors:

- The complexity and criticality of design;
- The risk associated with the design;
- Costs associated with design, development, or procurement;

- Commonality with other designs or subsystems.

CIs can vary in size and complexity from an aircraft, ship, or electronic system to a test meter or round of ammunition [1], and there may also be a hierarchy of CIs—that is, CIs may contain lower level CIs. In our aircraft example, the aircraft engine subsystem may be a single CI (if, for example, an unmodified OTS engine from a third-party supplier is selected), but the same subsystem may consist of many smaller CIs if modifications to an OTS engine are required or if a new engine is being developed for this aircraft.

So now we know the system requirements for the aircraft system as they are grouped functionally in the populated RBS at the system level. We also have decided what the aircraft system is to consist of physically at the subsystem level in terms of subsystems and CIs. What is left to do is to join the dots and relate the functional requirements to the physical aircraft CI architecture. This is the process of requirements allocation as performed in preliminary design and represents the translation from a functional to a physical architecture. Requirements allocation tells us exactly what each CI needs to do in terms of function and performance.

A proposed allocation is shown in Figure 3.6. On the left-hand side, we see the requirements of the aircraft system (functional representation), and on the right-hand side are listed the major subsystems, or CIs (physical representation). The two sides are related by the multitude of lines in the middle.

Figure 3.6 illustrates a simple method of showing functional to physical relationships, which becomes unreadable very quickly. A more useful way of showing the same information is via a matrix, sometimes called an allocation matrix or traceability matrix. The information in Figure 3.6 is shown in Figure 3.7 in the form of a simple allocation matrix for our aircraft system.

In Figure 3.7, the RBS functional requirements (representing the functional design of the aircraft system) are shown on the left-hand side of the matrix, and the physical design is shown across the top of the matrix as a series of subsystems or CIs described previously. The designers of the individual CIs can clearly identify what requirements and functions have been allocated to their CIs. The designers derive a host of additional subsystem functions and requirements for their individual CIs to ensure that the CIs are not only capable of meeting the assigned system-level and subsystem-level requirements but that they can also be integrated as needed. These additional requirements are known as derived requirements, since they have been derived from the need to satisfy some higher level requirement (an example of requirements flow down) or from the need to be integrated together (for example, interface requirements). The higher level requirements and the derived requirements for each CI will form the basis of the technical specification for that CI. The exact form of the CI technical specification depends on the design alternative selected to produce the

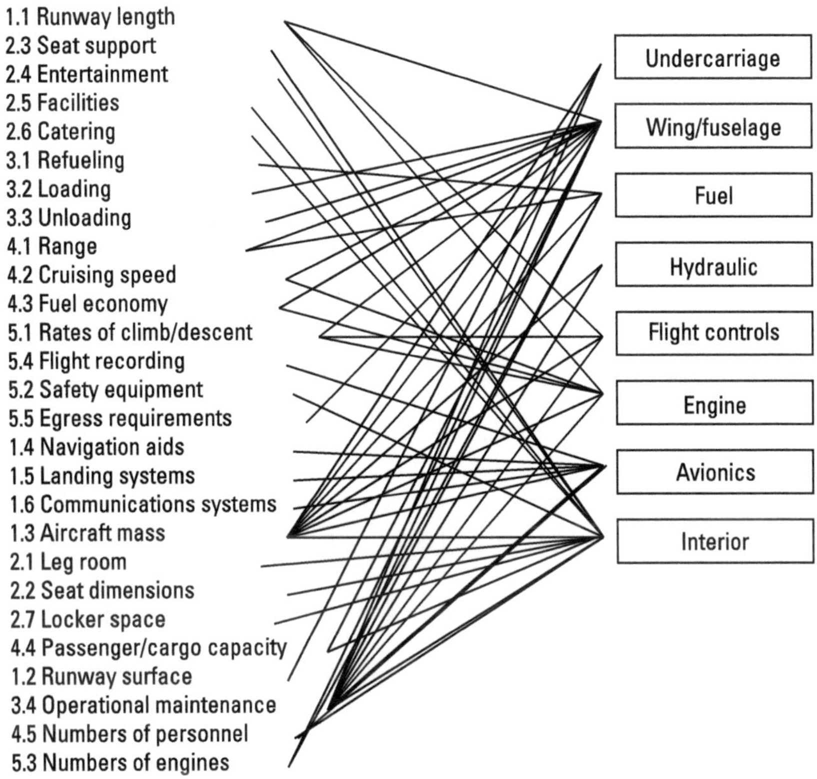

Figure 3.6 Simple example of allocation.

CI, but for developmental CIs the description of design called a development specification.

The allocation matrix concept is also a very powerful tool when changes to either the requirements or physical design are proposed. The impact of making changes to particular requirements or changes to specific CIs can be determined very quickly.

From the allocation matrix it is clear that some system-level requirements can naturally be assigned at a higher level to more than one physical CI, which means that the system-level requirement must be decomposed further before it can be allocated. In these cases, the collection of CIs is jointly responsible for satisfying the higher level requirement. It is worth investigating some examples to explain why this is often the case.

Example 3.3: Engine CI

Mass is a good example of a system-level requirement that has an impact on the design of all the CIs. In the aircraft example, we are limited in mass

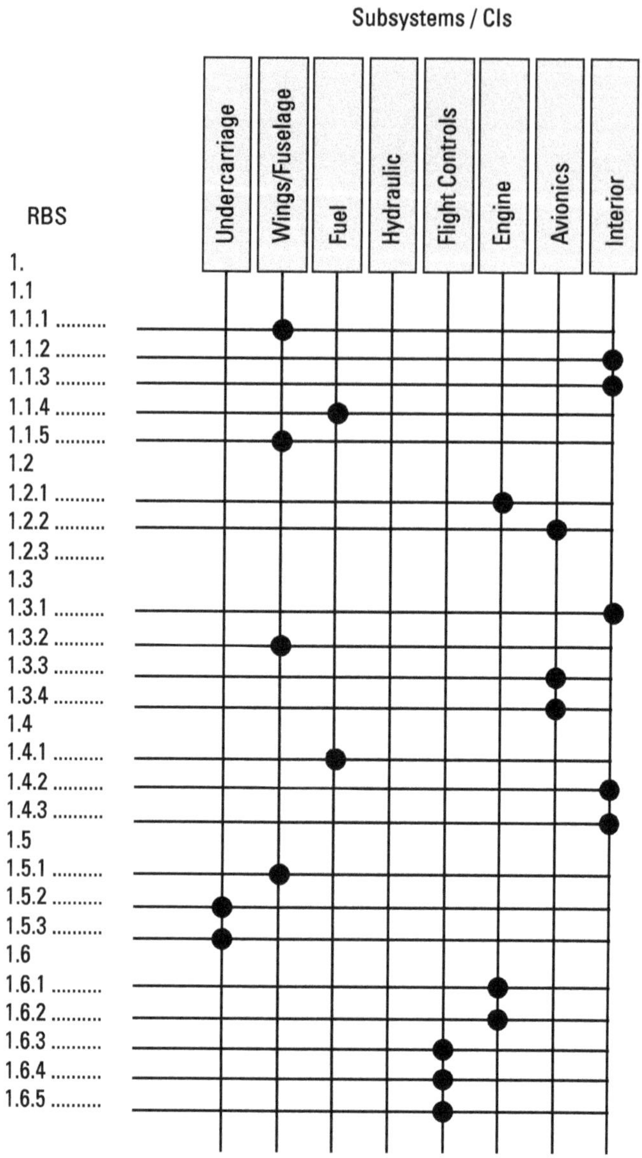

Figure 3.7 A simple example of an allocation matrix showing allocation of RBS functional requirements to physical subsystems/CIs for the aircraft system.

due to the requirement to operate from Class X airports. Since each CI contributes to the entire mass of the aircraft, the system-level mass requirement needs to be decomposed/derived, and a portion allocated to each CI. This allocation will result in derived CI mass requirements that sum to the system-

level requirement. For example, the Engine CI may be allocated 15% of the maximum entire aircraft mass meaning that the design of the engines must be achieved within that allocation to ensure that the entire aircraft mass is acceptable. To understand what else they must achieve with their engine design, the designers refer (vertically) to the allocation matrix and identify that their engine design must be capable of the following:

- *Propelling the aircraft sufficiently to takeoff on a runway of a certain length;*
- *Achieving a certain fuel efficiency;*
- *Achieving and maintaining a minimum cruising speed;*
- *Enabling the aircraft to achieve minimum rates of climb and descent;*
- *Being maintainable and reliable enough to allow operational-level maintenance to be performed within the required amount of time.*

From these requirements that have been allocated to the engine CI, the designers derive requirements specific to the engine. For the engine CI, designers may arrive at the type of engine to be used, the required levels of thrust and other performance imperatives, air intake requirements, electrical requirements, mounting and interface requirements, among others. These derived requirements and allocated system-level requirements form part of the development specification for the engine CI that, when complete, is known as the allocated baseline for that CI.

The process of allocation has ensured that each functional requirement (shown on the left of the allocation matrix in Figure 3.7) has been assigned to at least one CI for satisfaction. By working horizontally in the matrix, we are effectively working from functional to physical design (top down). By working vertically in the matrix, we can appreciate the responsibility each CI has for achieving allocated requirements (bottom up). Bottom-up analysis shows how the functional requirements of the system drive the design of each of the CIs (not the other way around).

It is now appropriate to ensure that the system-level TPMs determined during conceptual design have been decomposed/derived and allocated to the various subsystems and CIs in the system design along with the other requirements. This process will help determine specific qualitative and quantitative parameters that must be met by each of the CIs in order that the system-level TPM is met. The decomposed/derived and allocated TPMs assigned to the CIs are included in the specifications for the CI, bearing in mind that these CIs will either be procured (off-the-shelf), designed from scratch, or a combination of both. Either way, the CIs must contribute to meeting the TPMs determined during conceptual design and assigned during preliminary design.

It is important to note the advantages of the top-down design approach as borne out by the selection of design elements to meet the derived/decomposed requirements, including the TPMs assigned during preliminary design. The subsystems procured or designed for the system are done so with the system-level functional requirements and TPMs in mind. These subsystems are also procured or designed with a good understanding of the major interfacing and integration issues, which is discussed in Section 3.5. By following the top-down approach, the subsystems stand a very good chance of being compatible with one another and being able to integrate through defined interfaces to meet successfully the functional requirements of the system.

3.4 RBS Versus WBS

Now that the idea of grouping and allocating functional design requirements to physical design CIs has been explained, it is worth revisiting the relationship between the well-known project management tool called the WBS and the systems engineering process. The previous examples have shown the process of establishing the functional design of the aircraft system and the eventual translation of that functional design to physical design. The functional design was articulated in the RBS, and its relationship to the physical design was illustrated graphically using an allocation matrix. The concept of a WBS is not far removed from the idea of the allocation matrix. The WBS records the work packages and products necessary to produce the aircraft system. The work packages and products necessarily include all the CIs listed in the allocation matrix example. To that end, there is a relationship between the WBS and the RBS that is not dissimilar to the relationship between the CI list and the RBS shown in Figure 3.7. However, the WBS must capture much more than the list of CIs. The WBS needs to capture all the work required to design, develop, integrate, record/document, produce, and test these CIs. The breakdown of the work and products in the WBS is based largely on project management imperatives to manage scope effectively, but it will also drive cost and schedule management.

Example 3.4: Example WBS for the Aircraft System

An example of an aircraft system WBS is shown in Figure 3.8 (based on [2]), including an illustration of the relative location of the subsystems/CIs (from the allocation matrix).

Note that although the CIs do make up a significant portion of the WBS, there is a substantial amount of additional work contained in the WBS. The CIs map onto the system-level functions in the contracted SyRS;

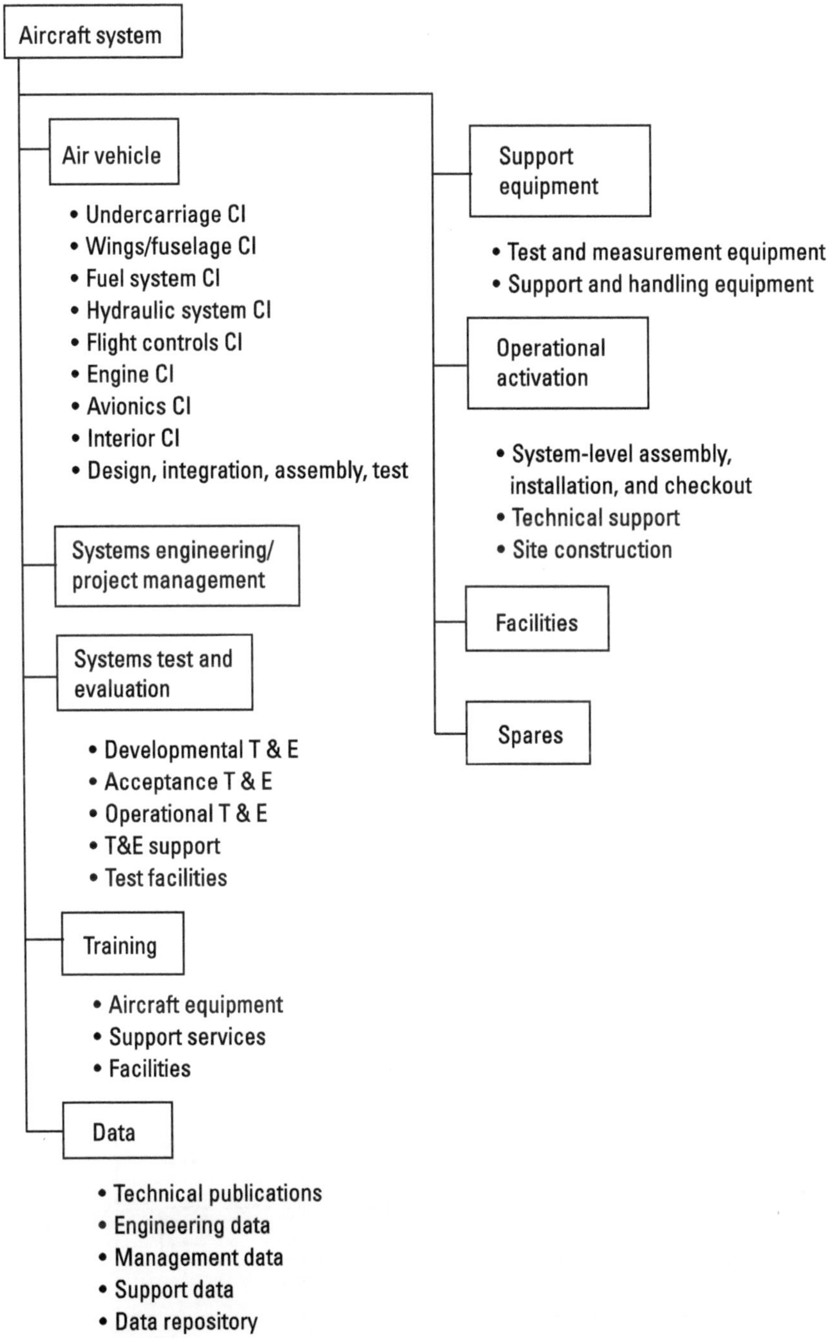

Figure 3.8 Aircraft system WBS.

the remaining work in the WBS maps onto the contracted SOW. Note also that the WBS is similar in structure to the RBS, although the WBS is organized physically, not functionally.

3.5 P3. Conduct Interface Identification and Design

During the selection of the CIs, the interfaces between the CIs must be identified, as illustrated in Figure 3.9.

Identification of interfaces is a critical part of preliminary design, as interfaces not only determine successful operation of the system once integrated, but they also place additional limitations and requirements on the design of the individual CIs. Interfaces must be accounted for, because, at the very least, they will require work that must be costed and scheduled.

The task of integrating different system elements begins with identifying and defining the interfaces that need to exist between the different CIs and between the CIs and systems that are external to the system of interest. Traditionally, this has been managed through the interface control document (ICD), which contains sufficient details to define the interfaces between the different subsystems and CIs and with external systems. Information relating to these interfaces may come from the definition of external interfaces and as the

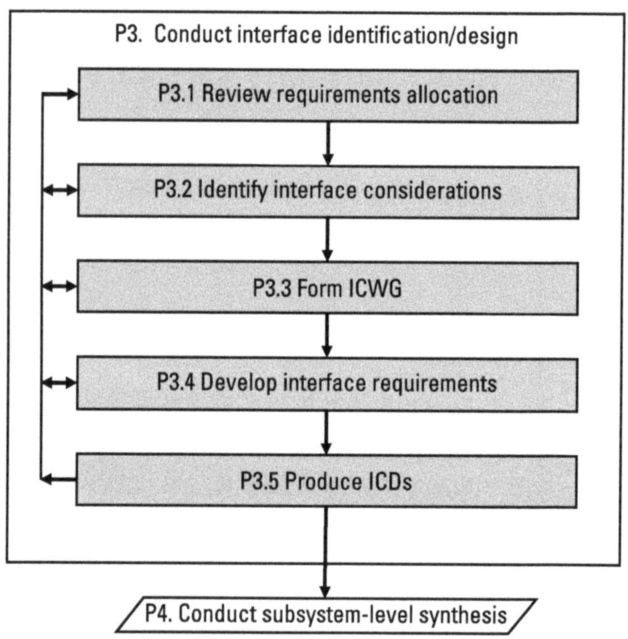

Figure 3.9 Interface identification and design.

functional flow block diagrams produced as part of the requirements analysis effort in the conceptual and preliminary design phases. Again, as with all systems engineering artifacts, the ICD can be text-based, data-based, or a model.

The ICD contains various types of information depending on the nature of the interface. Some of the different types of interfaces that exist are described as follows. (Note that to define completely some interfaces, an ICD may require information on more than one type of interface.)

- *Physical interfaces:* Physical interfaces are normally the most obvious of interfaces because they need to exist in real terms. They may be pipes through which a fluid or a gas flows, or they may be a fiber-optic cable passing digital information between two computer systems. In defining the physical interface, factors such as mechanical connection type and specification, physical size of the interface (such as diameter of pipe), and layout of the interface are defined. Physical interfaces normally coexist with one of the other types of interfaces.
- *Electronic interfaces:* An electronic interface is the name given to the flow of electronic signals (analog or digital) between two points. These signals may flow over a physical interface such as a cable or fiber, in which case the physical cable or fiber interface is also defined. Alternatively, the interface may make use of short-range RF transmission such as Wi-Fi or Bluetooth or longer range RF transmission such as microwave or satellite. The characteristics of the data flow between the two points need to be completely defined. The information requirements for this definition depend on the nature of the interface but may include decisions and constraints associated with modulation method, frequency, data format, data rates, and so on.
- *Electrical interfaces:* Electrical interfaces are normally associated with a physical interface of some sort such as a power cable or data cable. The type of information required in this sort of interface will include voltage levels, voltage type (e.g., 110V, 60 Hz), acceptable tolerances, and any associated factors such as fault condition requirements. In the case of a data cable, pin assignments and descriptions may be required. If there are any electromagnetic compatibility issues relating to the interface such as shielding and earthing, these requirements will also be defined.
- *Hydraulic/pneumatic interfaces:* Hydraulic and pneumatic interfaces are the mechanical equivalent of the electronic or electrical interfaces already described. A physical interface definition will coexist with this type of interface (such as a pipe or hose and associated connectors). Additional

information such as flow rate, pressure, temperature, and fluid type will be required to define completely the hydraulic and pneumatic interfaces.

- *Software interfaces:* Software interfaces refer to the passing of data between different computer software CIs. This may occur within one piece of hardware using internal hardware buses or it may be between two different pieces of hardware. If the latter is the case, a physical interface definition is also required to define the physical link between the two pieces of hardware. Even if software CIs are hosted on the same piece of computer hardware, there is a need to identify and define the software interface properly. Software interfaces require information such as expected inputs, generated outputs, format of the messages and any protocols used during the transmission. Protocol definition includes data rate, error detection and correction, start and stop codes, and so on.
- *Environmental interfaces:* Environmental interfaces are sometimes not considered an interface, but they do form the interface between the system and its external environment. In some cases, the environment provides the system with important information, such as air data in the case of an aircraft. The environment may also provide the system with challenging interfaces such as structural (vibration, shock, acoustics), thermal (ranges of temperatures), electromagnetic (flux densities), and radiation (various types). All environmental interfaces need to be considered and documented to ensure that the most important integration of them all (a system with its operational environment) is successful.

Example 3.5: Interfaces and CI Design

Example 3.3 looked at the design of the engine CI using the allocated set of functional requirements to drive the derivation of additional requirements. The allocated requirements are not the only set of requirements that will have an impact on the design of the engine.

The engine must interface with several other CIs including the Fuel CI. An interface between the Fuel and Engine CIs will be identified and documented during preliminary design (see also Section 5.8 on integration management). The documentation of this interface includes details of the physical interface between the two CIs and describes critical performance factors such as maximum flow rates, pressures, and physical means of connection. Both the Engine CI designers and the Fuel System CI designers have specific requirements from the interface. To that end, an interface control working group (ICWG) consisting of representatives from both CIs may be formed to participate in the design of the interface.

Characteristics of the interface design (such as flow rates, fuel pressure and temperature, and physical means of connection) form additional constraints on the design of both the Fuel and Engine CIs. The constraints produce a number of derived requirements that must be included in the relevant specifications for the CIs.

3.6 Subsystem-Level Synthesis and Evaluation

Once the allocation of requirements has been performed, it is time to continue the familiar loop of analysis, synthesis, and evaluation and concentrate on synthesis. Subsystem-level synthesis results in a preliminary design comprising hardware, software, personnel, facilities, data, and so on to meet the subsystem-level requirements. Figure 3.10 illustrates the process.

3.6.1 P4.1 Review Requirements Allocation and Interfaces

The allocation process that has grouped similar functions together into logical groupings has already completed the initial analysis. Designers responsible for each CI have continued the analysis to an additional level of detail necessary to support the subsystem-level synthesis. Technical specifications and ICDs have

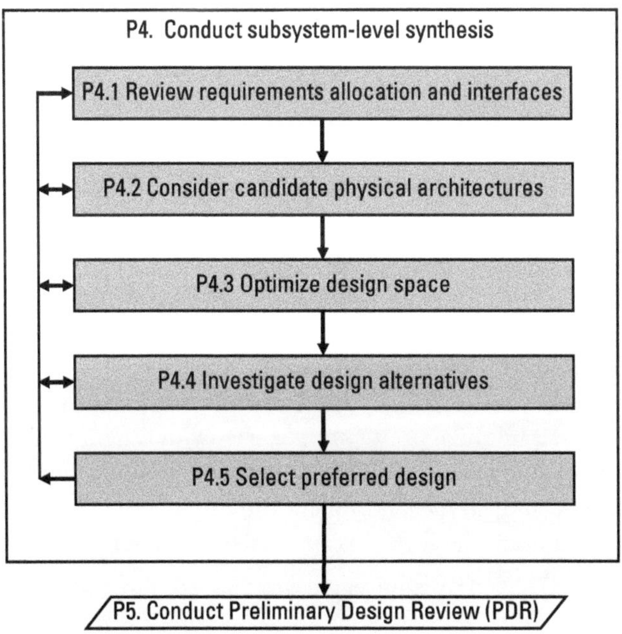

Figure 3.10 Subsystem-level synthesis.

been produced to detail the functions, performance, and interface requirements for each CI. The initial synthesis is conducted to determine a preliminary design that satisfies the functional groupings using one of the three broad design options available. As a first step in the synthesis process, the specification and ICDs associated with each CI should be reviewed, since it is these specifications that must be satisfied by subsystem synthesis.

3.6.2 P4.2 Consider Candidate Physical Architectures

There are three broad design alternatives available including off-the-shelf (OTS), modified-OTS, and developmental items.

- *OTS:* OTS equipment, as the name suggests, refers to pieces of equipment that are readily available. Commercial-OTS (COTS) equipment is commercially available; such equipment in the military environment is often called military-OTS (MOTS), and similarly, government-OTS (GOTS) in the government environment. If the OTS equipment meets or exceeds the requirements detailed in the development specification, there may be advantages to using OTS equipment. The most obvious advantages are that the equipment is likely to be readily available (no delay); it is also almost certainly cheaper than the alternative, and maintenance and support will probably be already in place for the item. The disadvantages often associated with OTS include problems with form, fit, or function, including inappropriate size, shape, mass, or color, or excessive or insufficient functionality. OTS items will ideally be described by a product specification delivered with the equipment, but in all likelihood the documentation set available will be substantially less detailed than a classic product specification. The documentation does, however, need to be detailed enough to support tasks such as integration, maintenance, and support. The need to qualify new pieces of OTS equipment in the system's operational environment may also affect the decision regarding the use of OTS. Qualification is a time-consuming and expensive exercise and may reduce considerably the benefits associated with the use of OTS equipment. For example, before using an OTS radio system in our aircraft example, the radio system would need to be proven or qualified using a set of standard test procedures to ensure that the radio could survive and function under the temperature, vibration, shock, and humidity extremes associated with aircraft operations.
- *Modified-OTS:* OTS equipment that meets most requirements contained in the development specification for a CI may be modified to better suit the needs of the designer. Similar advantages will exist with modified-OTS as for OTS; however some maintenance and support

may be voided if the item is modified. Another point to consider is the effort involved in modifying the item. This effort is easily underestimated, and the modification process can take a lot longer and cost a lot more money than initially estimated. Modified-OTS equipment may be described in a number of ways. One way is to describe the original OTS equipment using the standard OTS product specification and to describe the necessary modifications to the OTS item with an OTS modification specification. An OTS modification specification may include a combination of process specifications to describe how to modify the OTS item and material specifications detailing the materials needed to carry out the modification. These specifications are described in more detail in Chapter 5.

- *Developmental items:* If suitable OTS equipment is not available, the designer may opt to design and develop the item "from the ground up" to meet the specific requirements and characteristics detailed in the relevant development specifications. An item developed from the ground up may end up matching the desired criteria precisely in terms of form, fit, and function; however the effort involved in this development will be significant. Production processes and systems will need to be established, and maintenance and support issues will also need to be considered and resolved when items have been designed specifically for the system. Hardware and software development specifications, and hardware and software product specifications will describe developmental items in progressively more detail. These documents will contain information on the specific item and its interfaces. Hardware items may then require assembly or fabrication process specifications and associated material specifications to complete the description.

Other factors that have an impact on the designer's decision to use OTS, modified-OTS, or specially developed items include the availability and stability of the current technology, the size of the market, supportability, and cost. This helps account for future life cycle issues associated with OTS items such as obsolescence if production of the OTS item ceases, and support and spares become increasingly difficult to source.

3.6.3 P4.3/P4.4 Optimize Design Space and Investigate Design Alternatives

During the selection of the different subsystems making up the system, it is important to remember that it is the system performance that is of vital importance, not the performance of the individual subsystems and CIs. The reason system performance needs to remain the focus of the preliminary design is that to achieve optimal system performance, the performance of the subsystems and CIs may need to be suboptimal. A simple example illustrates this concept.

Example 3.6: System/Subsystem Optimality

> *Assume that a traffic management system (Figure 3.11) consists of a set of road subsystems (AA, BB, CC, 11, 22, 33). For each road subsystem to operate optimally with maximum traffic flow, all traffic lights on that subsystem must remain green. While that may be a reasonable design decision in the context of any single subsystem, this situation will clearly result in suboptimal system-level performance because it is not possible for traffic to flow efficiently through the entire network if only one set of traffic lights remains green. Note that this example has been simplified to illustrate the concept of design space. Most design space problems are multidimensional, representing nonlinear problems that require multidisciplinary teams to solve.*

To cater to the concept of optimal system performance, trade-offs need to be considered during the selection of the preferred design approach. These trade-offs relate to the performance requirements of the different subsystems and that of the system and any design space available for trade-off purposes. This helps ensure that the most important system-level performance measures (as ranked during conceptual design) are met first followed by the less important parameters.

Example 3.7: Aircraft System Design Space

> *The aircraft system development has reached a stage where the design of the CIs can commence. This example investigates the design of the Interior CI and the actual passenger seat. The example demonstrates the concept of design space and its use during preliminary design (and subsequent design activities).*
>
> *We know that seat design is of critical importance to the class-leading comfort required for the aircraft system, and we can trace the seat dimensions*

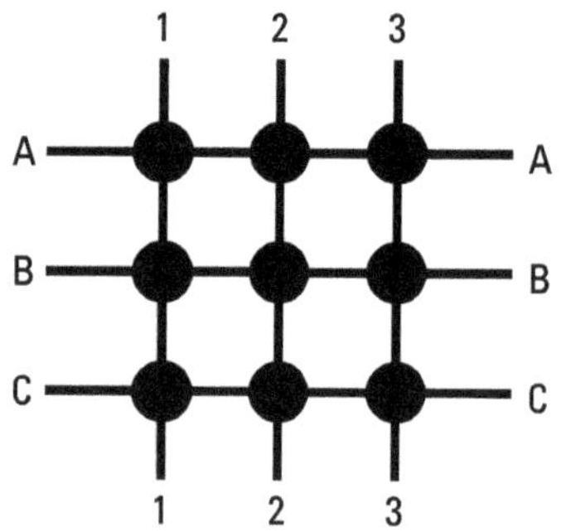

Figure 3.11 A road system comprising a number of subsystems.

back to the StRS. Let's say for this example that the minimum acceptable seat size is 576 in^2. Seat dimensions have been allocated to the Interior CI. When speaking to the designers responsible for the seat design, they explain that the number of passengers on board critically limits the possible seat dimensions. Naturally, the more passengers, the smaller each seat becomes for a given overall length allocated to the Interior CI. We also have the requirement for the number of passengers that the aircraft system must be capable of transporting. For the purposes of this example, the minimum acceptable passenger load is 150 people.

The relationship between seat dimensions and passenger load (as determined by the designers of the Interior CI) is shown simplistically in Figure 3.12.

Based on Figure 3.12 alone, we feel confident that we can meet both requirements. In fact, we can either increase the number of passengers to 170 (exceeding the minimum requirement for 150) and still meet the seat dimension requirement, or we could meet the passenger requirement and provide each passenger with a much bigger seat, or we could choose somewhere in between. In short, we have some "design space." The commercial viability of the aircraft (from which the passenger capacity requirements were derived) may be deemed more important than exceeding passenger comfort levels. If this is the case, then the decision may be to maximize the number of passengers on the aircraft and opt for 170 passengers with 576 in^2 seats.

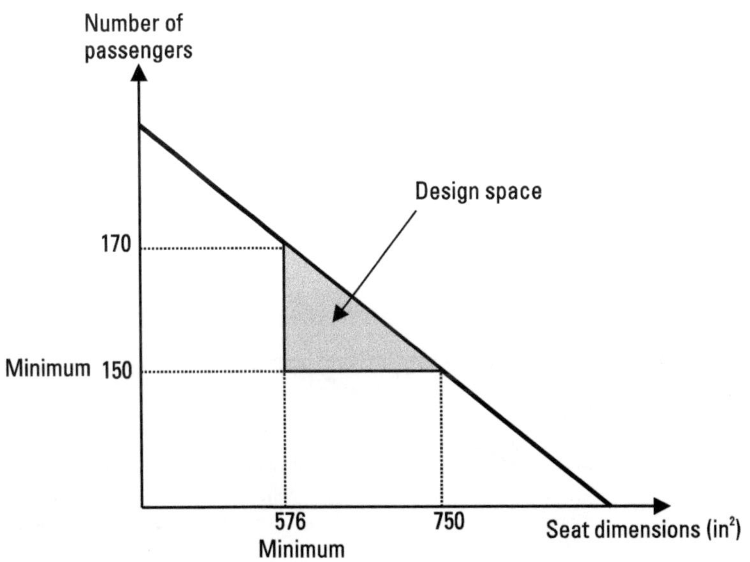

Figure 3.12 The design space represented by the relationship between the number of passengers and the seat dimensions.

Provided we keep within the design space available, we will meet or exceed all requirements. The aim in using design space is not to exceed some requirements at the expense of others, but rather to strike a balance and use the design space to at least satisfy all requirements. To use design space appropriately, however, the stakeholders must be consulted. Designers do not necessarily own the design space; the stakeholders do. In this example, the commercial stakeholders were given priority over the stakeholders vying for additional seat size.

We also know that, to meet the StRS requirement to operate from all Class X airfields in the world, a maximum aircraft mass constraint was determined. For this example, we assume that the aircraft mass is not to exceed 130,000 lbs. Mass was allocated to all CIs including the Interior CI. Figure 3.13 illustrates how passenger numbers affect the entire aircraft mass.

The design space relating to passenger numbers has now shrunk somewhat from that described in Figure 3.12. Based only on seat dimensions, the aircraft would be capable of carrying 170 passengers. However, when the entire aircraft mass is considered, only 160 passengers can be carried. The designers of the Interior CI will be happy, because they will now be able to exceed the minimum seat dimensions set for them (see Figure 3.12). Note also that the aircraft now has a minimum achievable entire mass of 120,000 lbs if we only carry the minimum acceptable number of passengers.

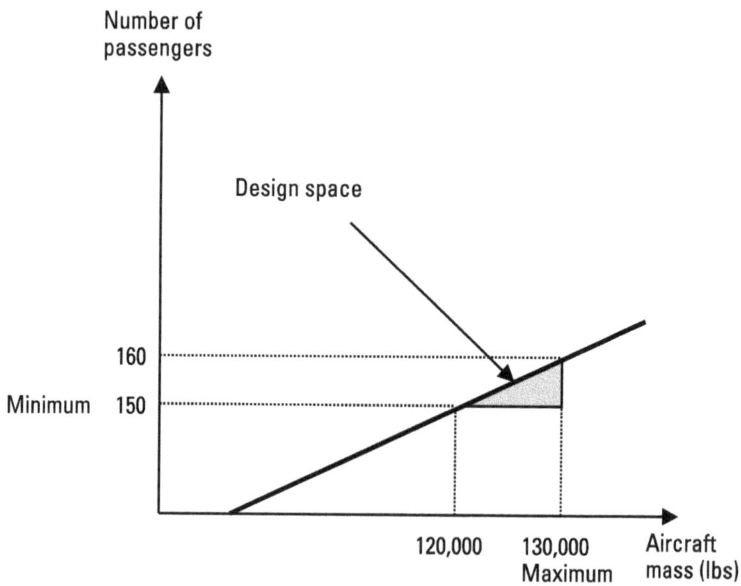

Figure 3.13 The design space represented by the relationship between the number of passengers and the total aircraft mass.

Often, however, this type of analysis reveals unavoidable conflicts between requirements and the design of individual CIs. For example, the designers of the Engine CI may identify an OTS engine that meets or exceeds all allocated and derived requirements, except for the minimum thrust requirement. The engine does not produce sufficient levels of thrust to support the allocated requirements of minimum runway length, cruising speed, and rates of climb/descent for a 130,000-lb aircraft. The designers are attracted to the OTS engine because it offers all the classic advantages of OTS design: reduced schedule, less cost and lower technical risk, and well-established logistics support. In addition to this information, the designers indicate that it is not technically feasible to provide the necessary levels of thrust from an aircraft engine using current technology for a 130,000-lb aircraft. The engine designers illustrate their case with Figure 3.14, showing thrust versus mass necessary to achieve the allocated requirements of runway length, cruising speed, and rates of climb/descent.

Figure 3.14 clearly shows that the engine will be incapable of supporting the performance requirements for an aircraft that exceeds 115,000 lbs. In isolation, this is not a problem because 130,000 lbs is, after all, a maximum and a 115,000-lb aircraft will obviously meet that requirement. However, when passenger load is revisited, it is seen from Figure 3.14 that the 115,000-lb upper limit is not sufficient to meet the minimum passenger load of 150. This is an important conflict in the requirements.

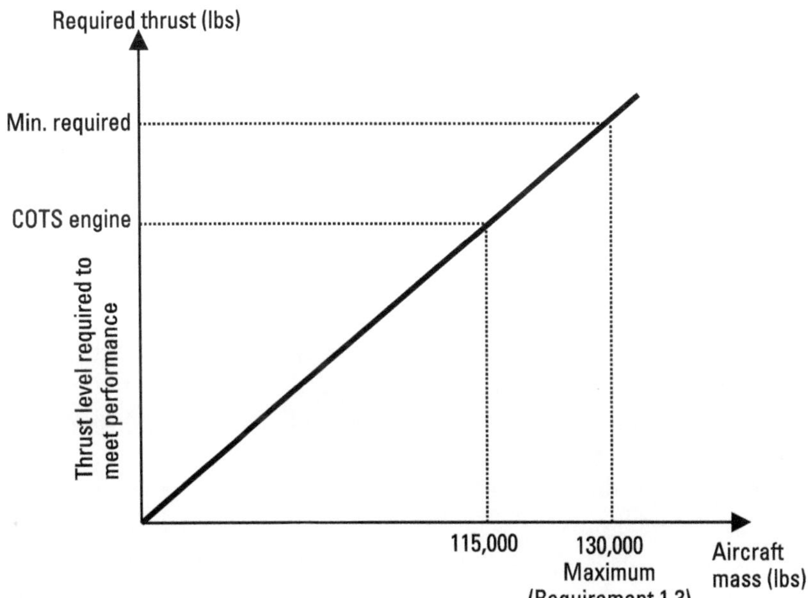

Figure 3.14 Revised power-versus-capacity function.

Traceability now becomes critical as the systems engineer traces the requirement for minimum passenger load back to the requirements in FBL (at the system level) from which that requirement was derived. The performance requirements came, in part, from regulatory requirements that are not negotiable. In this simple example, it is easy to remember that passenger load was related to commercial viability. The systems engineer must alert the owners (the relevant stakeholders) of this requirement as soon as the conflict becomes apparent, seeking guidance, clarification, and prioritization. Based on the feedback, design decisions can be made while keeping the owners informed. The system-level designers may be able to save some mass in other areas of design (by referring to the design space) to allow the minimum passenger load to be achieved while keeping the entire aircraft mass within the 115,000-lb limit set by the engine design.

This simple exercise has served to demonstrate the concept of design space and trade-offs to meet the original requirements of the system. The example has also demonstrated the concept of suboptimality where subsystem design must not take priority over achievement of system-level performance requirements.

Trade-offs within available design space needs to be applied to all the subsystems making up the system (as defined during the subsystem functional analysis and requirements allocation effort). Each subsystem affects different parts of the functionality and has different TPMs assigned to it. To that end, each unit has its own set of trade-offs and design spaces within which it must operate.

3.6.4 P4.5 Select Preferred Design

The purpose of allocation of requirements is to assign some performance guidelines to each of the subsystems and CIs making up the system to allow design engineers to select the subsystem that will be compatible with, and contribute to satisfying, the system-level requirements. This process is iterative, involving the familiar activities of analysis, synthesis, and evaluation. It may not be possible to meet all allocated requirements exactly on the first run through the design. The design (or synthesis) is evaluated to determine the level of success in meeting the requirements. Shortfalls are identified and the process must be repeated with different variables being traded off until as many of the allocated requirements as possible are met.

Once the preliminary design process is mature, the result is a preliminary architecture made up of various subsystems and their CIs of hardware and software designed and organized in such a way to meet a series of requirements that have been assigned to each component of the preliminary design.

Each subsystem has an individual description of its intended purpose and the requirements that it must be capable of meeting to fit into the system as a whole—called a specification (remember that a specification can be text-based, data-based, or a model, or some combination). These individual specifications must be detailed enough for a hardware engineer to design and build a piece of equipment (if that is deemed to be an appropriate approach) or for a software engineer to design and ultimately write executable software to perform the functionality, or for a logistics person to source a piece of OTS hardware or software to satisfy the requirements, and so on. Clearly, the level of definition for these individual specifications will vary. In the case of the hardware and software design options, these specifications are called development specifications. For the OTS hardware or software, the specifications can move straight to the product specification for the specific piece of OTS to be procured.

This allocation of requirements to system components and the development of detailed product specifications and development specifications is the ABL, the major product of preliminary design.

3.7 P5. Conduct PDR

As with all formal design reviews, the PDR is aimed at ensuring the adequacy of the preliminary design effort prior to allowing the design focus to shift onto detailed design. PDR is designed to assess the technical adequacy of the proposed solution in terms of technical risk and the likely satisfaction of the functional baseline. PDR also investigates the identification of subsystem interfaces and the compatibility of each of the CIs. The major PDR activities are illustrated in Figure 3.15.

As with all reviews, PDR must be conducted only when the acquirer is satisfied that the subsystem design has progressed to such an extent as to justify holding the design review. Prior to commencing PDR, the acquirer must be confident that the system architecture is in place at the subsystem level with all CIs identified and documented to an appropriate level of detail in development specifications. Interface identification and design should also be confirmed by investigating the state of the ICDs prior to the review, and evidence of ICWGs working effectively may be presented at the review. The acquirer also needs to be satisfied that other entry criteria such as financial and project management criteria have been satisfied.

The major task during the PDR is to investigate each of the CIs that have been determined during preliminary design. This process is aimed at verifying that all the overall system-level requirements are being met. In addition, the PDR investigates the results of the subsystem requirements analysis, the requirements allocation process, the various trade-off studies conducted during

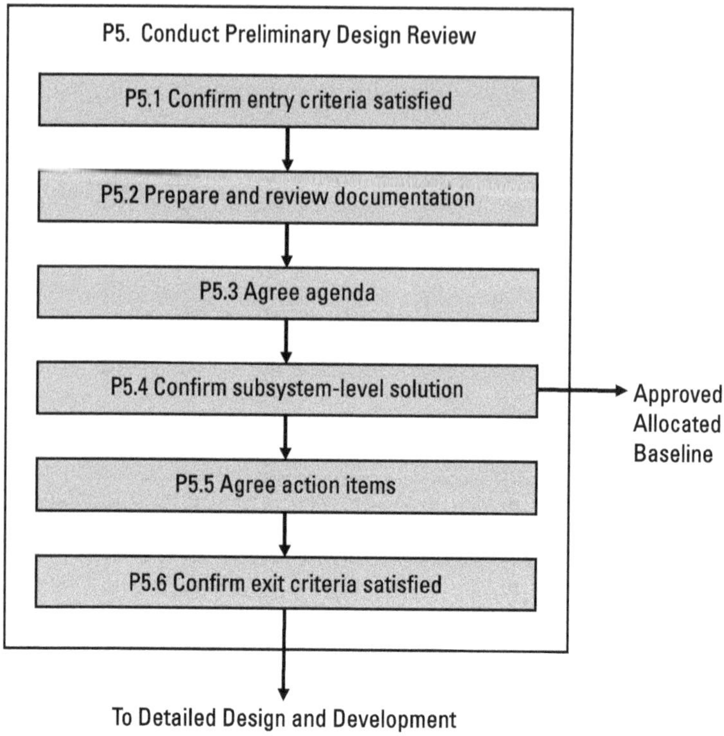

Figure 3.15 Major PDR activities.

the synthesis stage, and so on. Again, the results of these activities should be seen to address all requirements. Any deviations from the requirements are noted, and corrective action is determined and initiated.

At the conclusion of a successful PDR, the participants should be confident that all the FBL requirements have been addressed adequately by subsystem design. The functional and physical interfaces that need to be in place between the CIs, and between the CIs and the external environment, have been identified and documented ready for detailed design. The intended approach to detailed design should also be clear and acceptable, including the proposed use of OTS items and the scope of any detailed design and development effort. Participants should also be satisfied that the ABL is in a fit state to be used during detailed design and development. Any additional exit criteria listed against PDR should also have been adequately demonstrated. The successful conclusion of PDR results in the establishment of the ABL against which detailed design is subsequently authorized. Sometimes, an issue is raised during PDR that needs resolution prior to commencing detailed design and development. Either the PDR can be stopped until the issue is resolved, or a decision is made to continue with the PDR and the issue dealt with separately in a different forum.

Two approaches can be taken with respect to PDR. Either a separate PDR can be conducted for each of the major subsystems (or CIs) of the system, or a single PDR can be conducted covering multiple CIs. The approach taken should depend on the size and complexity of the system under investigation. The principal documentation reviewed and approved at PDR are the development specifications relating to each of the CIs and the ICDs describing internal and external interfaces. Preliminary product specifications (if available) may also be reviewed at PDR to ensure that the effort is on track, although these are often not available until deeper into the design process.

The aims of the PDR include the following:

- *Evaluation of the preliminary design:* PDR investigates preliminary design activities to ensure that they have been conducted thoroughly.
- *Approval of development specifications and ICDs:* The subsystems defined during preliminary design are documented using development specifications and ICDs. PDR reviews these documents and approves them prior to detailed design and development.
- *Assessment of traceability:* The review needs to ensure that all the requirements in the FBL have been adequately captured and addressed in the preliminary design. This is the concept of forward traceability introduced in Chapter 1. Backward traceability also needs to be confirmed to ensure that preliminary design does not include unnecessary gold-plating.
- *Establishment of the ABL:* The development specifications, ICDs, and top-level design developed during preliminary design form the ABL. This baseline is formally established at the conclusion of PDR and paves the way for the detailed design effort.

Other documents supporting the top-level design are reviewed during PDR if available. Examples of additional documentation include test plans, preliminary engineering drawings, and production plans.

Particular attention should be paid to high-risk areas, interface requirements, and system-level trade-off analyses. If appropriate, TPMs identified earlier in the system design effort should be discussed and progress traced during PDR.

3.8 Summary

This chapter discusses preliminary design, which starts with the initial FBL defined during conceptual design and continues with the effort of translating system-level requirements into design requirements for the physical subsystems that will be integrated to form the system. The activities conducted during the preliminary design effort include subsystem requirements analysis; subsystem requirements allocation; subsystem synthesis and evaluation; and establishment of an ABL. Responsibility for performing preliminary design normally rests with the developer; the acquirer's role becomes one of monitoring, reviewing, and supporting the developer's progress.

Now that the ABL is established at PDR, attention turns to detailed design and development, as discussed in Chapter 4.

Review Questions

1. Briefly describe the major activities and artifacts of preliminary design.
2. Briefly describe the P1. Conduct subsystem requirements analysis process within preliminary design.
3. Briefly describe the P2. Conduct requirements allocation process within preliminary design.
4. Briefly describe the four broad factors that guide the identification of configuration items (CIs).
5. Explain the relationship between an RBS and a WBS.
6. An allocation matrix shows the relationship between logical and physical design. Explain why this relationship is important to (a) the contractor responsible for design, and (b) the customer.
7. Describe the three broad design options (including their advantages and disadvantages) for CIs: OTS, modified-OTS, and developmental items.
8. Briefly describe the role of the ICD.
9. Briefly explain the role of an ICWG in the management of interfaces.
10. List and briefly describe the types of interfaces between subsystems.
11. Briefly explain what is meant by the notion of design space.
12. Briefly explain why a systems designer may have to force a subsystem designer to accept a suboptimal design space for their subsystem.
13. Briefly explain the role and conduct of the PDR.

Endnotes

[1] MIL-STD-481B, *Military Standard Configuration Control—Engineering Changes (Short Form), Deviations and Waivers,* US Department of Defense, Washington, DC, 15 July 1988.

[2] This example, which provides an excellent reference to WBS, is derived from that given in MIL-HDBK-881A, *Work Breakdown Structures for Defense Material Items,* Washington, DC: US Department of Defense, 2005.

4

Detailed Design and Development

4.1 Introduction

The detailed design and development activity continues the engineering effort and makes use of the FBL and ABL developed during conceptual design and preliminary design, respectively. The detailed design effort takes these definitions of the overall system (as in the FBL) and of the major subsystems and CIs (as in the ABL) and starts to consider the design of the developmental subsystem and components. The realization and recording of individual components are captured in the PBL.

A number of tasks are undertaken during detailed design and development. The major technical activities include the following:

- Describing the lower level components (hardware and software) making up the subsystems, including their interrelationships;
- Defining the characteristics of the above items through specifications and design data;
- Finalizing the design of all interfaces necessary to support system integration from the lowest levels up to the system level;
- Either procuring the items OTS, or designing them if they are unique to the system under development;
- Developing a prototype or engineering model of the final system by integrating the items (for the purpose of subsystem- and system-level test and evaluation);

- Redesigning and retesting (as required) aspects of the system prototype that failed to meet the minimum requirements;
- Conducting a design review (a CDR) to ensure that the design is ready for construction and production.

4.2 Detailed Design Requirements

Specific detailed design requirements must be derived from the various development and product specifications (derived during preliminary design) and through that process, related back to the system specification (developed during conceptual design). The content and intent of the specifications developed in detailed design and development are detailed further in Chapter 5. The detailed design specifications contain allocated requirements, further decomposed/derived requirements associated with TPMs, and associated characteristics that must be incorporated into the design of specific components. The specifications also necessarily contain descriptions of the requirements (functional, physical, and performance) for the interfaces relevant to the products, components, and assemblies. In addition, there may be a requirement to determine and specify the fabrication processes and materials needed to produce the detailed design. This is important because the materials out of which an item is fabricated and the fabrication processes will combine to define some of the function and performance characteristics of that item.

The specification process is an excellent example of top-down design process, which is responsible for establishing specific requirements at each level in the system's hierarchical design structure. The process evolves through the now familiar iterations of analysis, synthesis, and evaluation until the definition of all system components (recorded in the appropriate specifications) is complete. The collection of specifications that define the system's components combine to form the system's PBL.

4.3 Designing and Integrating System Elements

Through the process of system and subsystem requirements analysis and allocation, all elements of the system and their specific characteristics have been identified and recorded. The entire system has now been defined in terms of "whats." The design engineer now needs to use trade-off analyses and other design tools to determine the best way to deliver these "whats." That is, the "hows" need determination if they have not already been determined in an earlier phase.

4.3.1 Detailed Design Process

A typical detailed design process is very iterative in nature, is dominated by "reviews" and feedback at each stage, and makes use of a myriad of engineering design tools. The detailed design process is initiated by the completion of the previous phases in the acquisition process (conceptual and preliminary design). The SyRS and set of development specifications, provided by preliminary design, enable designers to determine the detailed design and definition of the subsystems and CIs (via the familiar analysis, synthesis, and evaluation process). These definitions are then reviewed to ensure that the definitions are complete and meet the overall system-level requirements. If the review is negative, the process is repeated. The detailed design process is shown in Figure 4.1.

The subsystem definitions are then further investigated and broken down into lower level portions of the design (including components, units, assemblies, software, subassemblies, and parts). Again, the review process ensures that this process is complete and accurate before the design effort focuses on producing the specifications detailed earlier to record the design effort. The system is

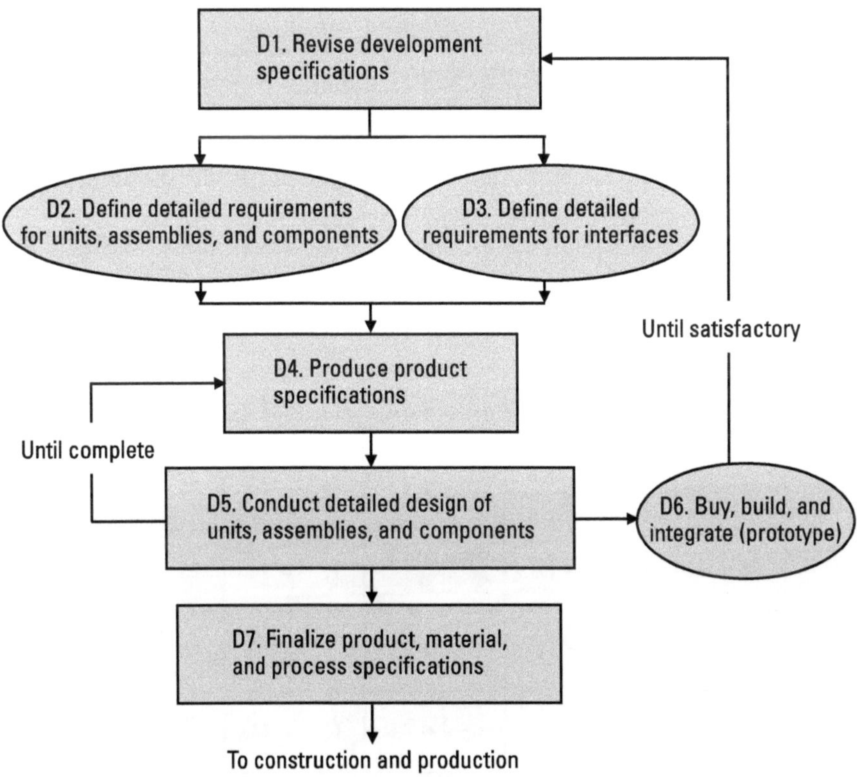

Figure 4.1 Detailed design and development process.

then procured (in the case of OTS items), modified, or developed (in the case of developmental items), and integrated as discussed in Section 4.3.2 to produce an initial prototype upon which system-level testing can take place. If the result of this testing is positive, the system can move into construction and/or production. If not, design changes or modifications will be recommended to address the discrepancies. Depending on the extent of the changes and modifications needed, the process returns to one of the previously detailed design stages.

4.3.2 Integration

Successful integration relies on accurate and complete interface definition during the preceding phases of the design. The system-level ICD and the corresponding individual interface descriptions contained in development and product specifications all play an important role in the integration task. In addition to adequate recording, cooperation between the various design groups (such as hardware and software) is vitally important, which drives the need for ICWGs to be formed. Effective ICWGs and interface discipline become more important when system components are being designed and developed in different geographical locations by different design teams and are only brought together during the integration stages.

An effective interface definition and design tool is known as the N2 diagram [1]. Although originally a software engineering tool, the diagrams are extremely useful in all types of interfaces. In simple terms, the diagram represents a matrix with system functions running down the diagonal. An example of a generic N2 diagram is illustrated in Figure 4.2.

In Figure 4.2, the interfaces are shown on either side of the diagonal; the interfaces are identified in both directions to take account of the fact that an

Function 1 (F1)	F1 → F2	F1 → F3	F1 → F4	F1 → F5
F2 → F1	**Function 2 (F2)**	F2 → F3	F2 → F4	F2 → F5
F3 → F1	F3 → F2	**Function 3 (F3)**	F3 → F4	F3 → F5
F4 → F1	F4 → F2	F4 → F3	**Function 4 (F4)**	F4 → F5
F5 → F1	F5 → F2	F5 → F3	F5 → F4	**Function 5 (F5)**

Figure 4.2 An example generic N2 diagram.

interface may represent an input to one function and an output from another. If no interfaces exist between certain functions, the relevant box in the diagram is left blank.

The N2 diagram can also use physical items in place of functions. During preliminary design, these physical items will be the CIs that have been identified, and the interfaces will represent interfaces between the CIs. As the design becomes more detailed during detailed design and development, these interfaces will be refined further. In this way, the N2 diagram can support the top-down identification and design of interfaces throughout the development life cycle from very high-level functional interface identification during conceptual design to very specific and detailed interface design during detailed design and development.

Example 4.1: Aircraft System—Example Partial N2 Diagrams

An example of a partial CI-level N2 diagram for the aircraft system is shown in Figure 4.3. Note the inclusion of external systems to allow the identification of interfaces with external systems and the environment.

Each of the interfaces identified in the N2 diagram is considered separately using the ICDs described in Chapter 3. ICDs can exist at different levels in the system hierarchy. For example, an ICD may be used to help manage

Avionics	• Throttle controls • Electrical power	Fuel pump control		HMI
Mechanical power	**Engine**		Mechanical power	
Inputs to fuel gauges	Fuel pressure/ supply	**Fuel**		Current quantity indication
Inputs to hydraulic gauges		Pressure to fuel pumps	**Hydraulic**	Current quantity indication
Landing and navigation inputs		Refueling equipment	Refilling equipment	**External systems**

Figure 4.3 Example of a partial N2 diagram for selected physical elements of our example aircraft system.

system-level interface definition during conceptual design. Similarly, lower-level ICDs will be established during preliminary design to record the requirements associated with subsystem and CI level interfaces. Detailed design of each interface is conducted during detailed design and development and recorded in the relevant product specification. For example, the design details of the interface to provide hydraulic pressure to the fuel pumps in the fuel system would be recorded in the interface design section of the hydraulic subsystem's specification. The fuel subsystem description would also note that it is expecting hydraulic pressure to be provided by the hydraulic subsystem across that interface.

Detailed design has now reached a stage where each CI has been designed completely, as have the interfaces between the CIs and between the CIs and external systems. The individual items can now be procured (in the case of OTS items), procured and modified (in the case of modified-OTS items), or constructed from the ground up (in the case of developmental items). The items can then be integrated in accordance with the overriding system architecture to form the prototype of the final system. Integration aims to combine lower level components into progressively higher level subsystems until the system prototype is complete. While the design process has been conducted top-down, the integration process is conducted bottom-up. At each stage of the integration, some form of integration testing is conducted to verify the successful integration against the appropriate requirement sets (that is, the PBL and ABL). Eventually, when systems integration is complete, testing can be conducted at the system level against the FBL. T&E, which has an important role in all phases of the systems engineering effort, is described in greater detail in Chapter 5.

The integration effort is summarized in Figure 4.4.

4.3.3 Some Detailed Design Aids

Detailed design is a challenging task, and fortunately, as technology advances, sophisticated design aids are emerging to help the system designer. These aids allow the designer to realize rapidly and inexpensively different design alterna-

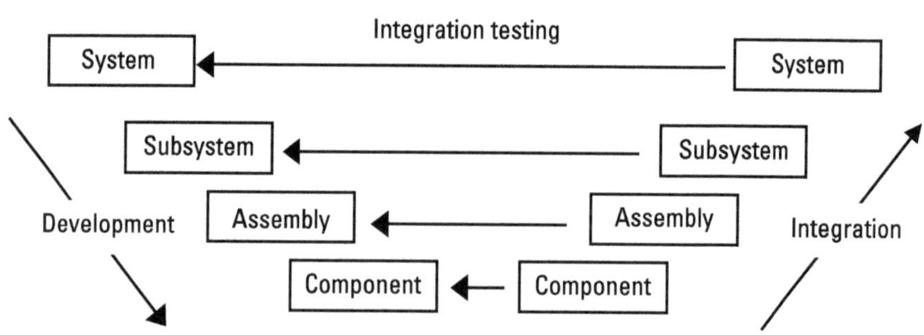

Figure 4.4 Top-down development and bottom-up integration process.

tives for the purpose of comparison and selection. The different alternatives can be traded off against one another to ensure that the most suitable design alternative is selected. This section provides some general examples of design aids used for hardware and software engineering. There is not sufficient space for the inclusion of more specific information.

Hardware engineering has benefited ever-increasing computing power available to the end user. Modern engineering design often makes use of sophisticated modeling and simulation tools that allow designers from various technical disciplines to develop designs concepts against allocated requirements sets, trade off different design alternatives, and carry out some level of requirements verification against allocated requirements, all within a virtual environment. Collectively, the integration of these tools and design processes into the detailed design and development stage is called model-based design (MBD) in part to differentiate it from MBSE (see Chapter 6), which can be used to support the systems engineering process particularly during conceptual and preliminary design and during system integration. MBD has many advantages over traditional engineering design techniques. It tends to be quicker and relatively cost-effective to allow designers to experiment with different design options in a safe environment without risking personnel safety, equipment damage, or damage to the Earth's environment. Modern MBD packages also ease recording burden by integrating with production systems and product life cycle management (PLM) systems to generate parts lists and the like for the final configuration of the design in the lead up to the production stage.

Modern detailed design tools mean that engineers can perform analysis, synthesis, and evaluation at each stage of the design process without necessarily having to create expensive and time-consuming physical engineering models and mock-ups of the elements they are designing. However, modeling and simulation tools have limitations, prompting the quip that all models are wrong, but some are useful [2]. Physical models and mock-ups are still often the simplest way for designers to verify aspects of physical layouts, experiment with human-machine interfaces, confirm access points, and so on. Physical models and mock-ups can also assist those in related disciplines, such as logisticians (to verify reliability and maintainability estimates) and production engineers (to confirm manufacturing processes) [3].

In terms of software development, there has been a similar trend in the growth of model-based tools and development environments that allow software engineers to design, test, and record software components of systems. A wide (and ever-increasing) range of development environments exists that allow software engineers to generate draft code automatically against allocated requirements sets and to develop effective software user interfaces and functional software for the purposes of design evaluation. Software development environments incorporate development tools, extensive software libraries, debuggers, editors, and compilers. Embedded software (used in microcontrollers, for example) can be developed, debugged, and compiled in a single environment, and the same environment usually incorporates the ability to simulate the behavior of the host hardware within which the embedded software will operate.

It is not necessary to go into detail about the design technologies available to modern designers. Suffice to say that technology is helping to make the detailed design process quicker and more likely to arrive at the correct answers. It is also tightening the integration between the systems engineering process (often now employing MBSE techniques) and the traditional detailed design process (leveraging MBD).

4.4 System Prototype Development

All the engineering design effort to this point has been focused on breaking the overall system-level requirements down to a detailed level, assigning these requirements to selected unit-level components (hardware, software, and so on), and then designing these individual elements. Analysis has been continually used to ensure that the overall system level requirements are being met by the design alternatives selected. Analysis forms one of the cornerstones of the systems engineering effort and is a vital part of the entire process. However, at this stage in the process, there is a need to verify the design concepts and system configuration using the actual system components arranged, combined, and integrated in a "final" state. The arrangement, combination, and integration of the components is sometimes called a system model or prototype. The system-level requirements can now be physically demonstrated using this model and a defined test and evaluation regime. This system-level test and evaluation is vital considering the concept of subsystem suboptimality introduced in Section 3.6.

Prototypes can vary from very complete and almost final versions of the system in terms of form, fit, and function through to models that only represent the functional characteristics of the final system.

Example 4.2: Aircraft System Prototypes

A prototype of our avionics system may be no more than a series of networked personal computers running a mature version of the avionics software suite, sometimes called an operational flight program (OFP). This network of computers may be operating in an office environment in the center of a major city.

Conversely, the same avionics system development may result in a prototype that consists of the final aircraft hardware mounted in a mock-up of the aircraft equipment bay and wired together using the actual wiring looms to be used in the aircraft installation. Actual aircraft instrumentation may be used and simulated parameters such as basic air data may be used to stimulate the model.

The type of prototype used will depend very much on the nature of the project and system. In the case of safety or mission critical systems, the system prototype must offer stakeholders high levels of confidence in the robustness of design. For example, aircraft avionics now has an impact on the very airworthiness of aircraft, so it is easy to justify very sophisticated prototypes to verify the correct performance of aircraft avionics software under a range of conditions prior to testing the avionics subsystem in an actual aircraft. On the other hand, a prototype designed to confirm the layout of the aircraft galley is likely to be quite a simple arrangement.

4.5 Detailed Design Reviews

As with each of the preceding design phases, a design review (or a series of reviews) is normally conducted at the conclusion of detailed design and development to ensure that the system design is complete prior to construction and/or production.

4.5.1 Hardware/Software Design Reviews

Design reviews are a natural part of the design process and will be conducted throughout the detailed design phase. These design reviews are likely to focus on individual items of software and hardware to ensure that the specific design approach meets allocated requirements. These reviews may investigate the results of design-related modeling and simulation drawings, early versions of software or hardware prototypes, and artifacts including specifications to thoroughly review the item under investigation.

Associated artifacts such as trade-off reports that justify and explain the design approach will also normally be investigated during these reviews. If component-level prototypes or models have been produced and tested, the results of the testing and any functional discrepancy rectifications will also be tabled.

All hardware/software design reviews should be completed prior to the major formal review of the phase, the CDR.

4.5.2 CDR

CDR is the final design review prior to confirmation that the design effort has been completed and that the system is ready to enter the planned construction and/or production stage. The result of the successful completion of CDR includes the establishment of the PBL, which marks the effective freezing of all design activity. Only discrepancies in the system design identified during testing result in further design activity following CDR. As with the PDR, CDR needs to be conducted on all the CIs of the system (hardware and software), particularly those that have had some level of design and development. Whether a separate CDR is conducted on each CI or a single system-level CDR is conducted (reviewing all CIs at once) depends on the size and complexity of the system under investigation. Another approach is to conduct CDRs of logical groups of CIs based on subsystems. For instance, in our aircraft example, we might conduct a CDR on all the CIs associated with the engine and conduct another CDR on the avionics CIs. This approach allows flexibility in scheduling, prevents a single CI holding up a major review, and supports sequencing to take account of design dependencies across CIs and subsystems.

Regardless of approach, CDR aims to confirm that the CIs under review can satisfy the functional and performance requirements allocated to them in the product specifications. CDR also confirms the compatibility of each CI with other CIs in the system. Developmental T&E activities will reach a critical stage and can be reported on, and the strategies, procedures and support required for a comprehensive acceptance test and evaluation can be confirmed. CDR is also the forum in which current plans for construction and production of the system are reviewed and reconfirmed.

A key aspect of CDR is the investigation of the TPMs identified during conceptual design whose progress has been tracked and monitored ever since. All TPMs should be either exceeding or meeting the necessary levels by CDR. By this mature stage in the design, there should be no surprises as far as TPMs are concerned. Departures from the accepted results should be noted and corrective action attempted, although at this late stage in the development process, significant departures are normally difficult to rectify.

As with all design reviews, a clear demonstration of readiness should be provided prior to commencing the review via the satisfaction of all entry criteria. All necessary artifacts must be received and reviewed in advance of the design review. Examples of artifacts that would be expected to be reviewed prior to CDR include revised interface descriptions such as ICDs; specifications such as development specifications and product specifications; T&E procedures and results; relevant technical data including assembly diagrams and drawings; installation drawings and schematics; software source code listings; and comments. Reviewers should be convinced that all necessary hardware or software reviews have been conducted prior to CDR. There is little point in conducting a system CDR before all CI-level reviews are complete. The system architecture and detailed design should be mature prior to commencing CDR, as CDR marks the effective freezing of design activities. It is likely that some integration and prototyping will have occurred since the last review (PDR), and CDR investigates and assesses the success of this effort as a key indicator of design maturity. If the design effort is ongoing, then the design is not ready for the CDR. This judgment will be possible following receipt and review of the artifacts already described. Once satisfied that the system design is ready for CDR, the review can commence.

For CDR to be considered a success, the reviewers need to be satisfied that the functional and performance requirements allocated to each CI will be satisfied by the proposed design for the CI presented at the review. This provides confidence that the requirements contained in the ABL and FBL stand a good chance of being achieved.

Out of CDR may flow the requirement to update the FBL or allocated baseline. The PBL is established by successful completion of CDR, and the system is then ready to move into construction and/or production in accordance with approved construction/production planning.

The timing of CDR within detailed design and development is critical. If the review is planned too early in the process, the design may be immature and may result in several postreview changes. Postreview changes have an impact on production planning and may introduce additional risks in the latter stages of the acquisition phase. Obviously, holding CDR too late in the project risks missing opportunities to rectify issues and may adversely affect the production schedule and may give rise to testing and delivery delays.

The aims of the CDR include the following:

- *Evaluation of the detailed design:* A great deal of effort has been put into establishing the detailed design of the system during this phase. This design is recorded in various forms including the mature product specifications and related engineering drawings. CDR should evaluate these artifacts and the detailed design to ensure that it adequately addresses the original requirements. Discrepancies raised during the PDR should be revisited to ensure that the rectifications are adequate. The review of TPM progress should be examined during CDR, and all TPMs should be well within expected bounds. Particular attention should be paid to any TPMs that were highlighted during PDR as being potential problems.
- *Determine readiness for production/construction:* In addition to the product specifications, ICDs, and drawings, CDR should evaluate the production plan and the quality assurance plans to ensure that the detailed design for the hardware items can progress to construction and/or production. The system prototype (if produced) provides an excellent tool for assessing readiness for production and/or construction.
- *Determine maturity of software:* Traditionally, software will enter the coding stage following CDR. (This can be considered the software equivalent of hardware fabrication.) The software product specifications (including

interface requirements and design) should be thoroughly reviewed prior to approving the software aspects of CDR. It should be noted that current software development processes allow software coding to begin long before construction and/or production. If this is the case, CDR should be used to monitor the progress of software development and investigate any software-related test and evaluation results.

- *Determine design compatibility:* CDR should investigate and confirm design compatibility of CIs and other aspects of the system such as external systems and interfaces. This requires a detailed review of all external and internal interfaces. The system prototype will provide an indication of design compatibility.
- *Establish the PBL:* The complete set of product specifications, once approved, will form the initial PBL for the system, which is required prior to the conclusion of detailed design and development. Functional and physical configuration audits are eventually conducted on the production system prior to the formal approval of the PBL for a system. These audits are discussed in Chapter 5.

4.6 Construction and/or Production

Once the PBL has been established following successful completion of detailed design and development, the system can move into construction and/or production. The term construction tends to be applied to infrastructure projects where a single (potentially massive) system is produced. For example, buildings, bridges, rail networks, roads, and tunnels are generally said to be constructed rather than produced. It is also applied in most cases to the process of creating software products—a piece of software is constructed and then replicated on host hardware. In contrast, production generally refers to a process where multiple copies of the system are going to be produced. Examples may include vehicles, aircraft, or domestic products like washing machines and dishwashers. Production involves some sort of production process that makes use of specified parts and raw or processed materials. It generally involves personnel, robots, and specialized machines carrying out repetitive tasks such as producing individual parts from raw or processed materials. In addition, production activities source OTS components and parts, assembling those low-level parts into higher assemblies using fabrication processes such as welding, fastening, soldering, assembly, and surface finishing, before continuing to integrate progressively higher level assemblies together to eventually produce the system. We refer here to this phase as construction and/or production, because one or both concepts will be applicable to the realization of our systems in most cases.

Although the construction and/or production process is presented here as a single life-cycle phase, it may commence gradually with what is sometimes called initial production or low-rate production. This initial phase can be used to prove that the production process is working and to allow verification of initial production versions of the system against the functional baseline. Adjustments may be made as required (to either the production process of aspects of the design or both) to improve the process and its output before moving toward a higher rate of production that is sometimes called full-rate production or serial production.

Regardless of the approach taken, production requirements need to be considered early in the acquisition phase to ensure that production risks are identified and addressed as early as possible. As with all other technical requirements, the earlier that production issues are identified in the acquisition, the easier they can be addressed. To that end, although construction and/or production is being considered following detailed design, production engineers should be working with the rest of the design team from the earliest possible stages of the system development to ensure that production and construction issues are appropriately addressed. In short, the design that flows out of the preceding stages needs to be producible.

Typical construction and production issues that need to be addressed and monitored throughout the entire systems engineering process include the following:

- Material availability (lead times), ordering, and handling;
- Fabrication requirements including tolerances;
- Process definition and process control;
- Assembly, inspection, and testing;
- Packaging, storage, and handling.

Strictly speaking, construction and/or production can be considered a specialty that supports the systems engineering processes. The systems engineering processes can be used before and during construction and/or production to support the construction and production activities. These processes include producibility analyses (ensuring that designs are in fact producible), consideration of producibility issues during trade-off studies, and the selection of preferred design options, and life cycle cost considerations that take the cost of production and construction into account.

Analysis-synthesis-evaluation is drawn upon again to support the construction and production effort; Figure 4.5 illustrates this process.

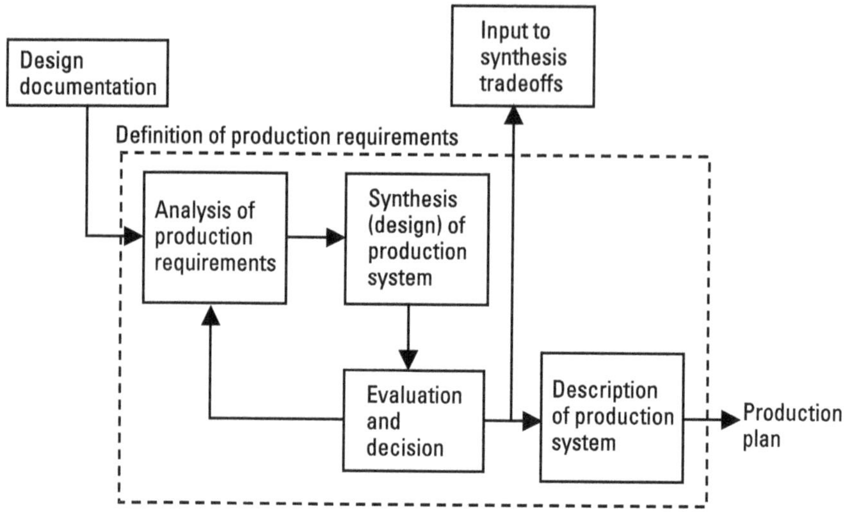

Figure 4.5 Systems engineering applied to planning for construction and/or production.

The output of the analysis-synthesis-evaluation process is a production plan that should be developed prior to the CDR and approved at that review. An approved production plan is required prior to entering construction and/or production. The content of the production plan varies widely with the specific characteristics of the system but should include the following considerations.

- *Resources:* The plan needs to consider the type and size of any plant or facility required to produce and construct the system. Personnel resources will include details of the skills needed by personnel and any training programs required.
- *Production engineering considerations:* Production and construction schedules will be needed, and the manufacturing methods and processes should be detailed. Special tooling and test equipment requirements must be specified. Any special facility-related requirements, such as storage and layout requirements, should be detailed. If any automated processes or manufacturing information systems are being used, these should be specified and described.
- *Materials and purchased parts:* A complete bill of materials should be compiled and listed. This will be available following the freezing of the detailed design artifacts. OTS or modified-OTS equipment should also be included in the plan so that all necessary items are covered. Any items involving long procurement lead times should be identified and ordered

in a timely manner. An inventory control system is probably necessary to track all the parts, and this system should be specified.

- *Management:* The standard management functions should also be outlined in the production plan, which should include details on process control, production testing procedures, subcontractor control, and problem prevention.
- *Logistics:* The plan should also detail any special logistics requirements such as packaging, handling, and storage requirements.

As systems are being realized by construction and/or production, they should be subjected to a range of acceptance tests and configuration audits. This is covered in more detail in Chapter 5. At this stage, it is sufficient to understand that objective evidence must be collected to verify that the system meets the requirements contained in the FBL. In addition to satisfying those stated requirements, configuration audits should be conducted to ensure that the system has been built in accordance with the approved process specifications, materials specifications, drawings, and part lists.

During construction and production, it may become clear that items of the system are not going to meet all the minimum requirements specified in the baselines. To cover this possibility, the configuration management process (see Chapter 5) uses a system of engineering changes and variances. Engineering changes are used where the departure from the specification is going to result in either a change in design or a change in requirement. If, however, the failure to meet the minimum requirements is temporary in nature, limited to only a small percentage of the supplies, or a very minor departure, a variance may be approved by the acquirer. Configuration management incorporating engineering changes and accommodating variances, a critical component of systems engineering management, is covered in detail in Chapter 5.

4.7 Operational Use and System Support

Once the system has passed the necessary acceptance testing and configuration audits, it is ready to enter operational service or use—the utilization phase. The major activities during the utilization phase include operational use, system support, and modifications. The systems engineering influence over these activities is relatively minor, although modifications and upgrades to the system over its life may present opportunities for systems engineering to contribute.

Configuration management must continue to play a role during modifications to the system to ensure that the system remains well-understood and recorded as it moves through the utilization phase. Differences between system descriptions and the physical system tend to creep in over time when complex systems are operated, maintained, and upgraded in an environment void of configuration management practices. These differences make maintenance and operation difficult and potentially dangerous. Sound artifacts support the need to understand the design and capability of the system, especially if there are slight differences in configuration across a fleet of systems (for example, an aircraft fleet). Modifications and upgrades made in accordance with configuration management practices enhance supportability and operations, but unmanaged changes to a system end up adversely affecting training, support, safety, and operations in the long run.

4.7.1 Modifications

Modifications and upgrades can help the system continue to meet operational and support requirements over an extended period. Failures of the system to meet current requirements may be identified as part of the failure reporting, analysis, and corrective action system (FRACAS) process described in Section 4.7.2, which may result in engineering changes to the system via the modification process. It may also be possible that modifications are required to rectify discrepancies with the performance of the system that were not identified during the acquisition phase. These discrepancies are often discovered during the operational T&E effort when the system is placed in its operational environment and exercised in its intended purpose. Modifications may also become necessary due to changing system-level requirements caused by a range of factors including a changing operational or support environment.

Regardless of their origins, modifications may have an impact the system performance and functionality. The degree of impact depends on the size and significance of the modification. As illustrated in Figure 4.6, significant modifications can be considered small system developments and may be executed using appropriate project management principles. That is to say that modifications are often considered projects that occur within the utilization phase of the system life cycle.

The following example illustrates how systems engineering methodologies may be employed to achieve modifications.

Example 4.3: New Engine for Our Aircraft System

Adding a new engine to our aircraft system would seem to be a simple process of changing a few details and revising the manufacturing process to produce the new variant. A simple modification, when viewed with respect to its impact on the original WBS, can take on more significant proportions. Example 3.7 illustrates the concept of design space and the balancing act required to ensure all system-level requirements were met. Changing a major part of the

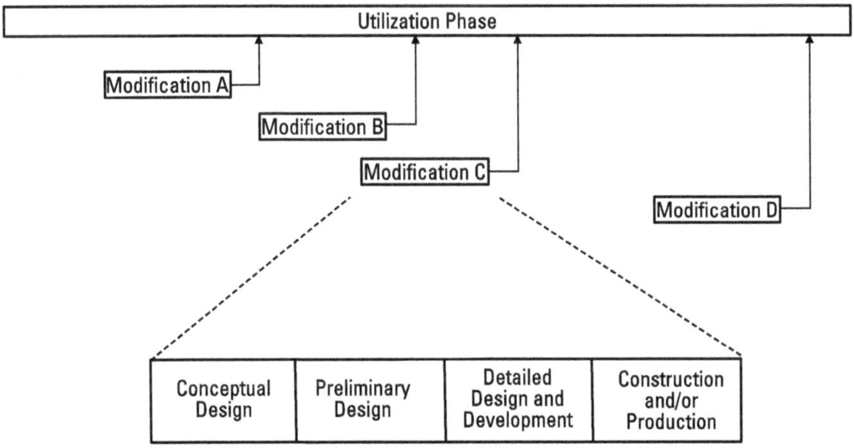

Figure 4.6 A number of system modifications may occur during the utilization phase; the systems engineering effort involved is highlighted for Modification C.

system can significantly upset the function and performance of the overall system. In Example 3.3, the Engine CI illustrated how interface requirements are born. With these two examples in mind, a change in the engine has a large number of flow-on effects to other CIs and ultimately to system performance.

Consider the Aircraft System WBS in Figure 3.8 and the impact that a simple engine modification may have on all of the elements in that WBS. The air vehicle element will be changed significantly, because the engine is a major component of the aircraft. It interfaces with a range of other CIs so changing an engine will also require changes to other CIs. T&E on the engine itself and any system-level functions affected by the engine will need to be rewritten and performed. Training will be affected as maintainers and operators will need to become familiar with the change. A new suite of technical and engineering data will be required to record the changes. Support equipment, spares, and facilities may also be impacted.

When the true scope of the modification is considered and investigated using the WBS, the modification starts to resemble a true system in its own right, and both systems engineering and project management methodologies can be employed successfully to achieve the modification.

Modifications have an impact on the original system; the degree of impact depends on the size and significance of the modification. Given that systems engineering can be employed to achieve modifications, systems engineering starts to play an increasingly important part in the utilization phase of the system life cycle as modifications become more prevalent. This gives rise to a revision of Figure 1.13 as shown in Figure 4.7.

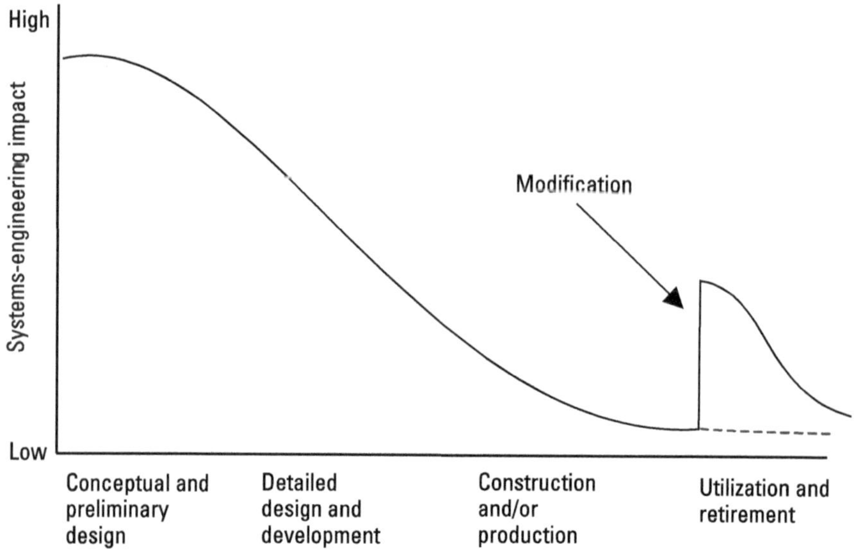

Figure 4.7 Effect of a modification on systems engineering impact.

System support also commences during the utilization phase to ensure that the system is maintained in operational order. The system support will be conducted in accordance with the approved support concept. Modifications to the system may be necessary to ensure that the system continues to be supportable and to help overcome the ever-present risk associated with technical obsolescence.

Example 4.4: Aircraft Modifications

Large aircraft systems such as military aircraft or airline aircraft are often designed to have a useful life of 30 to 50 years. During this time, changes in technology areas such as engines and avionics will continue at a rapid pace. It is common for the operators of these aircraft systems to conduct major modifications to their fleets. Modifications may include replacing engines with more modern equivalents and replacing aging avionics with more sophisticated solutions.

Such modifications have a great impact on our systems and may result in performance enhancements that allow additional tasks to be conducted; they may also allow continued operation in a changed environment. Although significant functionality enhancements often result from upgrade projects, the main goal of these projects may be to combat technical obsolescence to maintain access to spare parts and technical support. These programs help to ensure that systems remain maintainable and supportable through to the retirement phase.

4.7.2 FRACAS

Systems that operate in highly regulated environments (for example, defense and aerospace, transportation and automotive, medical devices, energy, and banking and communication) must comply with mandated standards associated with safety, reliability, and regulatory compliance. These standards often require the adoption of a formalized method and process to detect, report, analyze, and correct systemic failures. An example of such a process, a FRACAS is a closed-loop system designed to continually maintain visibility into system operation and support. The important difference between a maintenance system and a FRACAS is that the maintenance system rectifies failures, but the FRACAS attempts to rectify the cause of the failures (to prevent future failures). A successful FRACAS improves the reliability and maintainability of a system as it proceeds through its life cycle as systems normally become less reliable and more difficult to support and maintain as they age. The FRACAS is in place to help counter this natural decline in reliability and maintainability.

The recurring theme throughout this text is that problems discovered early in the systems engineering process are generally the easiest, cheapest, and quickest problems to rectify. The FRACAS, therefore, should be established as soon as possible, as it becomes critical as the components, subsystems, and system are produced, integrated, and tested. For OTS items, this could commence during preliminary design and for developmental items, during detailed design and development.

There are several industry standards that call for a process similar to a FRACAS [4]. MIL-HDBK-2155 [5] is a freely available reference that describes a FRACAS as having the following six steps:

- Failure reporting;
- Failure analysis;
- Failure verification;
- Corrective action;
- Failure report and close-out;
- Identification and control of failed items.

An adaptation of this process is shown in Figure 4.8.

The system under investigation is operated or tested in accordance with the standard procedures until a failure is noted. At this stage, the failure is

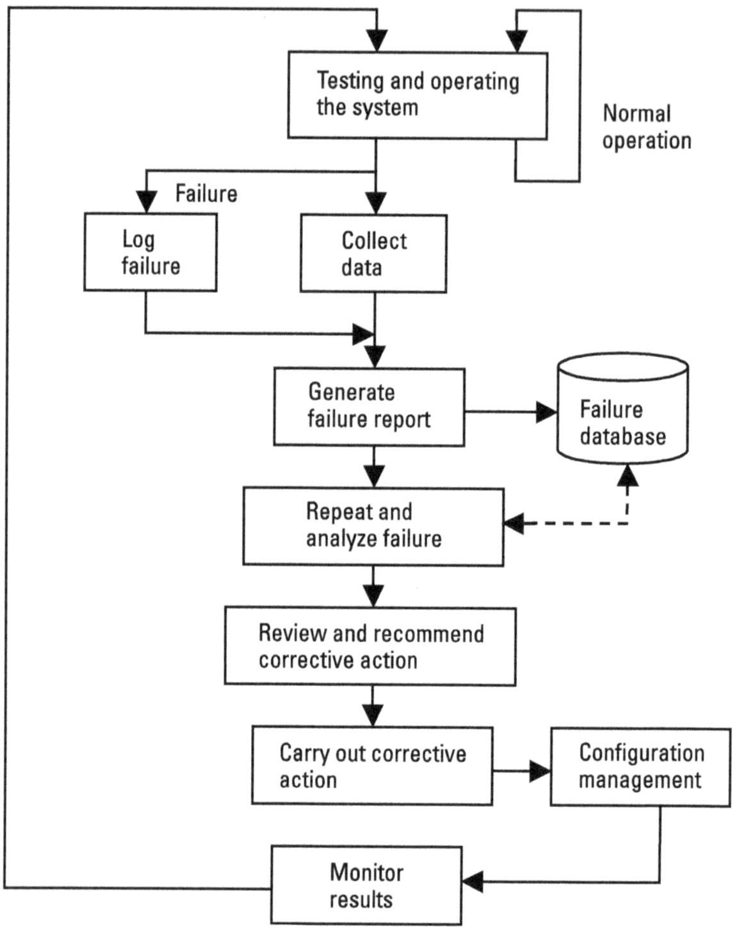

Figure 4.8 FRACAS process adapted from MIL-HDBK-2155.

logged and data collected to assist with the FRACAS process. The logging and data collection process needs to be recorded, and all personnel involved with the test or operation of the system should be educated in the process to ensure that it is applied rigorously and consistently. Testing personnel are normally well-versed in the FRACAS process, but sometimes failures go unreported during operations because operational personnel are not aware of the formal process in place. The success of the FRACAS relies on all failures being recorded, so it is critical that operational failures are included in the FRACAS; after all, the system spends most of its life in the utilization phase of the life cycle.

In the event of a failure, the data collected needs to be sufficiently detailed to support the FRACAS process. The data must describe the conditions under which the failure occurred, the symptoms of the failure, and quantitative descriptions of the result of the failure. The data supports the analysis process and allows investigators to replicate the failure if necessary and practical. To that end, it makes sense to involve the investigators in the design of the form that will be used to collect the data. A failure report is generated to record the failure and its associated data. Modern FRACAS processes support this operation with a database that simplifies the entry, storage, and retrieval of the information.

Once a failure report has been generated, responsibility for investigating the failure is delegated to the relevant personnel in accordance with the FRACAS process. The investigators are selected based on their experience and expertise and are charged with investigating the failure, analyzing the failure, and generating a recommended corrective action report. These personnel investigate the failure by using the reported data and searching the database for other similar or related failures. Under some circumstances, it may be necessary for the analysis process to repeat (and verify) the failure. Sometimes, however, this may not be possible, especially if the failure results in injury or death of personnel or substantial loss of equipment. The analysis process makes use of established failure analysis tools including failure modes effects, and criticality analysis (FMECAs), cause-and-effect analysis, and fault-tree analysis. Following the analysis and verification process, the investigators determine the possible forms of corrective action likely to rectify the cause of the failures and generate a corrective action report containing their recommendations. Corrective actions are not limited to hardware or software changes but can also include changes to the training of operators or maintenance personnel, or changes to operational or maintenance procedures.

The corrective action report is assessed and approved by a review board established in accordance with the FRACAS process. The review board is closely associated with the established configuration management process. If the recommended corrective actions involve class I or class II engineering changes, these are processed in accordance with the configuration management process. The impact of the change is monitored to confirm that the failure does not recur as the system continues operation or test.

4.8 Retirement Phase

The retirement phase is the final stage in the system life cycle. Functions associated with phaseout and disposal include transportation and handling, decomposition, and processing of the retiring system. A retirement concept should be developed as one of the LCs during the early stages of the acquisition phase of a system (see [6] for more details on design for retirement). If considered early, disposal and phaseout issues will form some of the criteria against which the system is designed ("design for disposability"). It is important, however, that system designers focus on retirement, rather than the more limiting issues of disposal—planning for disposal is important, but a system may retire from several life cycles before it is ultimately disposed of.

4.8.1 Retirement

As illustrated in the example in Figure 4.9, a system may be developed from conceptual design in an initial life cycle and then delivered, deployed, operated, and supported throughout that life cycle. If the organization that owns the system changes focus and moves away from the business area for which the system was acquired, the initial owners may sell the system to a second organization, which, having completed conceptual design and preliminary design in that life cycle, has decided to meet business requirements with an OTS solution, albeit secondhand. As illustrated in Figure 4.9, the system then enters a second life cycle, perhaps with several of its elements (the operating personnel, for example) replaced. When the second organization wishes to replace the system (perhaps because it has become costly to maintain), the system may be sold to a third organization that wishes to meet its systems design with a modified-OTS solution based on a refurbished secondhand system—again, several system elements may be added or removed and other elements added. As illustrated in Figure 4.9, the third organization acquires the system, refurbishes it, and then fields it in a third life cycle. In this example, Figure 4.9 illustrates that the system is uneconomical to repair at the end of the third life cycle and is disposed of as scrap.

Example 4.5: A-4 Skyhawks

> *The Royal Australian Navy (RAN) purchased A-4 Skyhawks directly from the manufacturer (McDonnell Douglas) in the 1960s. When these aircraft were retired from RAN service in the 1980s, some of them were sold to the*

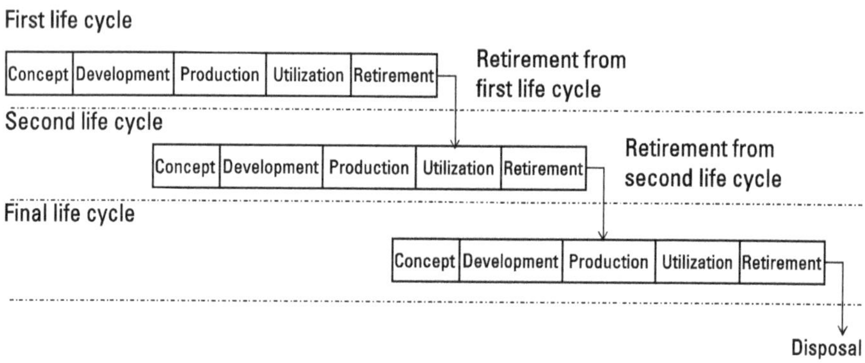

Figure 4.9 A system may be retired from a number of life cycles before final disposal at end of life.

Royal New Zealand Air Force (RNZAF) and began their second life cycle alongside other existing RNZAF A-4 aircraft. The RNZAF fleet of Skyhawks was eventually retired from service in the early part of the twenty-first century, and some were sold to a US flying school to begin their third (and possibly final) life cycle.

Designing for more than one life cycle is important, because, if designers assume that there is only one life cycle and design just for disposal of the system products at end of that life cycle, then that disposal method may well be the only option available at the end of that life cycle (or, at least, other options may have been constrained). Yet, for minimal environmental impact, system design should address the system's existence in as many life cycles as possible (on the basis that the longer a system exists, there is lower cost and lower environmental impact resulting from the need to develop replacement systems). A good system design should therefore not only achieve longevity of use in the first life cycle of the system but should aim for longevity of use in all life cycles until disposal is the only viable option remaining.

Retirement issues are often not considered until they are imminent. Very little is written about formal methods for the inclusion of retirement and disposal in early systems engineering activities. Perhaps as environmental and ecological constraints become more stringent, retirement and disposal will begin to play a more prominent part in the system life cycle. Indeed, environmental considerations such as emission control and material recycling are already playing an important role in constraining design, construction, operation, and disposal in many industries including the aircraft and motor vehicle industries.

When considering the retirement aspects of the system during its design, it is useful to undertake three broad tasks: to identify the reasons for potential retirement; to identify potential retirement methods for the system, which then allows conceptual designers to undertake a third task of identifying design issues that may arise from the consideration of each retirement method.

Some of the many reasons for potential retirement are described as follows.

- The system may have reached the end of its life and no longer be usable and/or supportable.
- The system may still largely be useful, but one or more critical components may be unusable or unsupportable.
- The system may still be perfectly useful, but it (or a significant subsystem of it) is being retired because of one of the following reasons:
  - The business need for the system may have disappeared (for example, the owner's business direction may change, or the owner may be retiring and winding up the business);

    - The business need is still valid, but the business owner is forced to retire the system (due to issues such as cash flow, a revoked operating license, scandal, public pressure, or disputes between business partners).
- The system (or one of its critical components) may be damaged beyond economical repair (by accident, natural disaster, act of war, or vandalism).

Once the reasons for potential retirement have been identified, each can be examined for potential retirement methods for the system (although not all may apply to all reasons for retirement). Potential retirement methods are listed as follows:

- Selling as a secondhand item;
- Trading for a replacement system;
- Redeployment as a training aid;
- Destruction and disposal as waste;
- Disassembly (i.e., systems are broken up and sold as working subsystems, assemblies, or components);
- Scrap (i.e., systems are broken up and sold for their value as scrap);
- Placement in storage (i.e., mothballed because none of the other methods are currently acceptable or tenable).

Once the reasons for potential retirement and the potential retirement methods have been identified with the stakeholders, designers can identify any design issues that may arise for retirement and disposal, such as the following:

- Observance of recycling regulations;
- Avoidance of the use of potentially hazardous/toxic materials;
- Environmental impact of disposal;
- Cost of disposal/destruction/uninstallation;
- Cost of refurbishment for resale or trade-in;
- Desirability of salvage or material recovery for resale;
- Cost of transportation for retirement or disposal;
- Need for specialized personnel/equipment required for retirement or disposal;

- Observance of any caveats on disposal (such as might be placed on military equipment by international arms trade agreements, or might be placed on high technology by national sanctions);
- Time taken to arrange disposal;
- Availability of design data and drawings to support disposal.

4.8.2 Disposal

Although designing for disposability may seem to be an unusual priority, consider the following examples of items that may be difficult and expensive to dispose:

- *Radioactive material:* Although many projects would not consider themselves a large user of nuclear material, common items such as infrared detection systems do contain radioactive material, and disposal raises a number of environmental issues.
- *Cryptographic material:* Many modern organizations make use of cryptographic material to protect information. Examples include military, government, and financial institutions. Although the storage and use of the cryptographic material is well-documented in various procedures, the disposal of items such as hard disk drives and floppy disks that have contained this information is not a simple matter. Discarding the devices will leave the organization open to compromise and a suitable disposal method must be formulated.
- *Normal building material:* Some materials considered normal have major impact during disposal. Consider the environmental and health concerns surrounding asbestos. The disposal costs of buildings containing asbestos are extremely high. It is most likely that an alternative product would have been used in most buildings where asbestos is now a problem if its nature had been known during preliminary design.

In reality, retirement and disposal are often not considered until they are imminent. Very little is written about formal methods for the inclusion of disposal and phase-out considerations in the early systems engineering activities. Perhaps as environmental and ecological constraints become more stringent, phaseout and disposal will begin to play a more prominent part in the system life cycle. Indeed, environmental considerations such as emission control and material recycling are already playing an important role in constraining design, construction, operation, and disposal in industries such as the aircraft and motor vehicle industries.

4.9 Summary

Detailed design and development continues the development effort of FBL and ABL in shaping the developmental subsystem and components. The realization and recording of individual components are captured in the PBL, which is confirmed at CDR; then, the system is ready to enter the construction and/or production stage.

Once the system has passed the necessary acceptance testing, it is ready to enter operational service or use—the utilization phase. The major activities during the utilization phase include operational use, system support, and modifications.

Modifications and upgrades can help the system continue to meet operational and support requirements over an extended period, but at some point, the system will no longer be able to meet its original purpose and will be retired and then disposed of at the end of its life.

In terms of the framework we introduced in Chapter 1, we have now finished our coverage of the systems engineering processes. Chapter 5 considers the broad topic of systems engineering management and details the major associated activities. Subsequently, Chapter 6 introduces some of the more common and popular systems engineering tools and explains their application to engineering management and processes. Finally, Chapter 7 explains the interrelationship between the systems engineering effort and other closely related disciplines and discusses the relationship between systems engineering methodologies and a number of acquisition methods.

Review Questions

1. Explain the role of system prototyping (including examples of different types) used during detailed design and development.
2. Describe the major review conducted at the conclusion of detailed design and development.
3. Explain why integration is the ultimate test of interface identification and design.
4. Explain the importance of some form of pr-production process (including a production readiness review) prior to finalizing the design and commencing production.
5. Detail the main activities associated with the utilization phase of a system's life cycle.
6. Name some possible reasons for modifications to the system during the utilization phase.

7. Explain, using an example, why systems engineering could be applied during a modification exercise.
8. Briefly describe the reasons for potential retirement of a system.
9. Briefly describe the potential methods for retirement of a system.
10. Briefly describe the design issues that might arise from consideration of system retirement.

Endnotes

[1] Walden, D. D, et al. (eds)., *Systems Engineering Handbook: A Guide for System Lifecycle Processes and Activities,* INCOSE-TP-2003-002-05, Hoboken, NJ: John Wiley & Sons, 2023.

[2] Box, G. E. P., and N. R. Draper, *Empirical Model-Building and Response Surfaces,* New York: John Wiley & Sons, 1987.

[3] See the following for further detail:

Blanchard, B., *Logistics Engineering and Management,* Englewood Cliffs, NJ: Prentice Hall, 1992.

[4] Examples of commercially available standards that either explicitly or implicitly require a FRACAS-like process include the following:

RTCA DO-178C Software Considerations in Airborne Systems and Equipment Certification.

RTCA DO-254 Design Assurance Guidance for Airborne Electronic Hardware SAE ARP5580 Recommended Failure Modes and Effects Analysis (FMEA) Practices For Non-Automobile Applications.

ISO 9001:2015 Quality Management Systems—Requirements ASD/AIA, S5000F International Specification for In-service Data Feedback, Part 1/2, 2021.

[5] MIL-HDBK-2155, *Military Standard Failure Reporting, Analysis and Corrective Action Systems,* US Department of Defense, Washington, DC, 1995.

[6] Ryan, M., "Design for Retirement," *Journal of Cleaner Production,* Vol. 70, 2014, pp. 203–210.

5

Systems Engineering Management

5.1 Introduction

Systems engineering management is responsible for planning and directing the systems engineering effort, monitoring and reporting on that effort to the appropriate areas, and reviewing and auditing the effort at critical stages in the system life cycle. This chapter considers the major systems engineering management elements: technical review and audit management, T&E, technical risk management, configuration management, the use of specifications and standards, and systems engineering management planning.

Systems engineering management is vitally important, but it also adds to the scope of work that needs to be managed by the project manager. Each of the topics discussed in this chapter is typically found within contractual SOWs and therefore needs to find its way into project WBSs and then into cost and schedule estimates. In this way, project managers are aware of not only the technical engineering work discussed in Chapters 2 to 4 but also the work and project deliverables associated with managing that technical effort. Additionally, care must be taken to tailor the systems engineering management activities to suit the project; otherwise, effort will be wasted—even worse, additional risk will be created by overmanagement.

5.2 Technical Reviews and Audits

As discussed in Chapters 1 to 4, the technical review and audit considerations of a project can be quite extensive, and the management of these issues helps ensure that the overall aim of reducing technical risks and enhancing confidence in the design process is achieved. So far, we have considered the major reviews and audits in each stage of the acquisition phase. This section discusses the management considerations of technical reviews and audits.

5.2.1 The Role of Technical Reviews and Audits

Technical reviews and audits provide both the customer/acquirer and contractor/supplier with a measure of progress toward the goal of successfully introducing a system into service; they also reduce the technical risks associated with system development. Reviews and audits achieve this by doing the following:

- Providing a formal evaluation of the design maturity;
- Measuring and reporting on planned and actual performance;
- Clarifying and prioritizing design requirements;
- Evaluating and establishing the system baselines at discrete points in the design process;
- Providing an effective means of formal communications between the stakeholders;
- Recording design decisions and rationales for later reference.

Technical reviews and audits are a vital part of the overall systems engineering effort and can range from formal and structured reviews associated with a critical system milestone to less formal reviews involving only a few people discussing a minor component of the overall system. Regardless of the level of the review, all reviews aim to determine the ability of the design in question to meet the relevant technical requirements. Reviews, which tend to become more detailed as the design progresses, cover all aspects of the engineering design effort (for example, software, hardware, and integration). The customer/acquirer normally specifies the required technical reviews and audits and their timing in the contractual/agreement artifacts (based on their initial expectations tempered by the nature of the successful tenderer's proposed approach to the technical program). Once agreed on, the contract/agreement specifies the length, duration, location, content, conduct, and attendees for each review. Based on these requirements, the supplier is better equipped to estimate the cost and schedule associated with the reviews. A full set of reviews in a large project can represent a sizeable portion of both time and cost. The contractor/supplier organization may also decide to include additional internal reviews to determine further the technical adequacy of the design.

The number of reviews required and their respective scope depends on the complexity and size of the system in question—so the reviews and audits must be tailored accordingly. The technical risk associated with the system also has an impact on the number of reviews scheduled. For example, a system that makes heavy use of unmodified OTS equipment may be seen as a low-risk project compared to a pure developmental project. Fewer lower level reviews would be expected for the OTS system than for the developmental project. In the case of modified OTS equipment, formal reviews may be limited in scope to cover the modifications.

Reviews and audits can provide significant assurance that system development is progressing technically. However, these reviews and audits also take considerable time and effort to plan, prepare, and conduct. Also, while reviews and audits are being conducted, attention may be diverted away from the design and development effort. It is important that the costs and benefits associated with conducting technical reviews and audits are considered fully prior to planning and mandating the reviews and audits to ensure that an appropriate level of review and audit is placed on the system development effort.

Inadequate review and audit may add risk to the system development by allowing potential design problems to go unchecked. The usual tendency, however, is to try to make the review and audit requirements bulletproof by overreviewing the project. Unnecessary reviews and audits increase the cost and schedule of the system without adding any additional value in the form of technical risk reduction. The additional cost and schedule are invariably ultimately passed onto the acquirer.

The scheduling of reviews and audits is also very important. Reviews held too early in the design process are unable to determine technical adequacy of the design, because the design is immature and subject to change. Reviews held too late in the process will miss the opportunity to avoid costly and time-consuming rectifications if the design proves inadequate. It is also useful to relate technical reviews and audits to documentation release [1]. For example, the PDR may be scheduled to coincide with the release of the hardware and software development specifications. Obviously, as the design progresses, reviews and audits require hardware and software products as well as documentation, but the concept has merit, because design documentation tends to be a strong indicator of design maturity.

To this point in the book, we have introduced the major formal reviews and audits referred to in MIL-STD-499B [2]. Although this standard was never released, it provides an excellent source of reference for formal technical reviews and audits and their intended purpose, and it is freely available. Another freely available reference source for technical reviews and audits is MIL-STD-1521B [3], which was canceled when MIL-STD-499B was released in draft. MIL-STD-1521B provides a detailed set of entry and exit criteria from each of the reviews and audits listed in this chapter. Several commercially available reference sources also describe the need for technical reviews and audits as part of a systems engineering process [4]. Where system size and complexity warrant it, configuration audits may be considered as a separate topic from technical reviews. MIL-STD-973 [5], MIL-HDBK-61B [6], and EIA-649C [7] provide comprehensive coverages of configuration audits, including the role that audits play in the overall configuration management process.

As mentioned earlier, not all reviews are applicable to all systems. To that end, some reviews and audits can be tailored out of the requirements to suit the individual requirements of each system.

5.2.2 Technical Review and Audit Management

As well as specifying the reviews and audits required in the contract/agreement, the acquirer should also specify how the reviews and audits are to be conducted. The requirement for agenda, data packages, and minute-taking will help in making the reviews and audits as effective as possible.

Both the acquirer and supplier are active participants in reviews and audits. Both parties need to prepare prior to the event to ensure that the attendees have the appropriate expertise to contribute to the areas being covered in the review. The attendees must be given adequate time to review the agendas and data packages prior to the review. Each person should have a role in the review, and participants should not consider the review as an educational experience. Observers may, however, attend the reviews as a convenient way for new staff to become familiar with the system and subsystem design.

Decisions, action items, and agreements should be noted in the minutes that are distributed following the meeting. Each action item should also be assigned to an individual who is responsible for following the action through to fruition by an agreed on date. Unassigned action items invariably remain unactioned.

At the conclusion of the review, the acquirer team should meet to determine the level of satisfaction with the meeting. Successful technical reviews are a good indication of project status and schedule. Significant milestone payments are often tied to the successful completion of design reviews. Technical reviews and audits are, therefore, an important part of both the systems engineering and the project management effort.

There must be a willingness to make difficult decisions in the event of unsatisfactory technical reviews and audits—which may result in the need to revisit some or all the preceding effort. The adage that bad news does not improve with time certainly applies to inadequate design. While there may be pressure to meet progress and schedule targets, it is often better to stop work, go back a little in the process and get it right before continuing. Succumbing to the pressure to continue is rarely in the interests of the project.

The technical review and audit effort is principally managed via a Technical Reviews and Audits Plan (TRAP). The TRAP documents all formal reviews, detailing the entry criteria that must be met prior to the commencement of the review or audit and the exit criteria that must be demonstrated prior to approval of the review or audit. The TRAP normally lists the design documents that need to be produced for evaluation during the relevant review. The TRAP is normally drafted during conceptual design and should be finalized and approved prior to any contract being signed.

5.3 System T&E

T&E is a systems engineering management function that ensures that a coordinated and consistent testing effort is applied to the system for the entire system life cycle. By coordinating the T&E from a system perspective, the focus and emphasis of the testing can be varied with different life cycle phases without compromising the entire T&E effort. T&E, which applies throughout the entire system life cycle, involves both the acquirer and supplier.

A thorough evaluation of a system involves validating the system against the original acquirer requirements. Obviously, this validation cannot be completed until the entire system has been designed, developed, and constructed and then operated in the intended operational environment by operational personnel. The aim of T&E is to test and evaluate the system progressively as it passes through the acquisition and utilization phases to avoid costly and time-consuming modifications to the system design late in the life cycle. Progressive T&E could be considered a risk mitigation measure as it provides a high degree of confidence early in the system life cycle that the design will ultimately perform as required.

Testing can be an expensive and time-consuming exercise that requires the use of specialized personnel, test equipment, and facilities. A formal plan is normally put into place very early in the Acquisition Phase to manage the system T&E effort. This plan, which can be called the Test and Evaluation Master Plan (TEMP), forms a major deliverable from the systems engineering effort. The TEMP is discussed in more detail later in this section.

The entire systems engineering process aims to produce a system that is both verified against the requirements generated during the systems engineering process and validated against the original needs, goals, and objectives that initiated the system development in the first case. Often these two associated aims are combined into the term V&V, which ensure that we have "built the system right" but that we have also "built the right system." A well-managed approach to T&E aims to support the delivery of a system that is both verified and validated. Other systems engineering management functions, including technical reviews and configuration audits, also support the overall objectives for verification and validation.

There are three major categories of T&E that are roughly applied to coincide with the acquisition phase, the transition between the acquisition and utilization phases, and the utilization phase. They are described as follows.

- *DT&E:* DT&E refers to the T&E activities undertaken during the acquisition phase of the system life cycle to support the design and development effort and to verify that the system's elements are being developed in accordance with the relevant element specification. DT&E activities may also occur during the utilization phase to support activities such as modification development.
- *AT&E:* As DT&E completion approaches, AT&E activities become increasingly relevant. AT&E represents the formal acceptance testing conducted on the system to enable the acquirer to accept the system from the supplier by verifying that the system meets the requirements outlined for it in the contracted SyRS. AT&E effectively forms the boundary or transition between the acquisition phase and the utilization phase. Unlike DT&E and OT&E, AT&E tends to be a discrete testing activity (with a defined start and a defined end).
- *OT&E:* OT&E focuses on the functional or operational testing of the capability system and its components, conducted under realistic operational conditions, including being operated by operational personnel and with all the other elements of capability. OT&E is normally conducted for a period following acceptance of the system by the acquirer, although limited OT&E activities are possible during acquisition, especially when a long production cycle means that some systems have been accepted prior to other systems being produced, or when concept demonstrators are being used. OT&E validates that the system meets the BNR/SNR as encapsulated particularly in the StRS and continues to do so throughout operational use and system support.

Figure 5.1 illustrates the role and timing of each type of testing.

This section describes the three forms of T&E and, where possible, details the specific test activities associated with each of the activities within the acquisition and utilization phases, as described in Chapters 2 to 4.

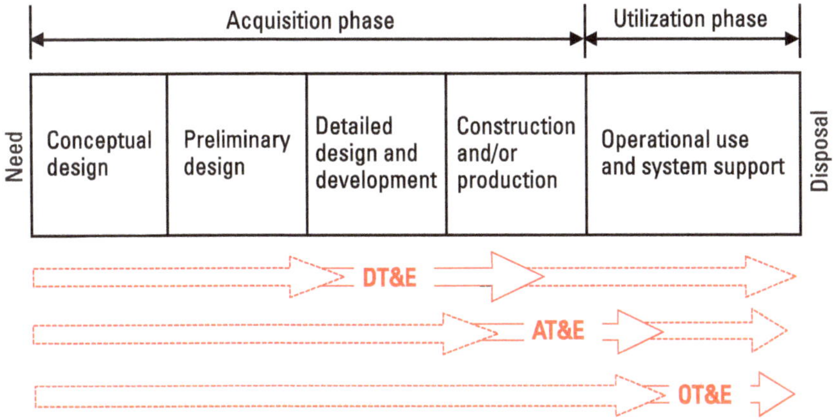

Figure 5.1 Timing of DT&E, AT&E, and OT&E.

5.3.1 DT&E

As the name suggests, DT&E encompasses the verification activities conducted throughout the acquisition phase of the system life cycle and aims to highlight deficiencies in the system design as early as possible. As we have seen several times, the earlier that problems are identified in the system life cycle, the cheaper and simpler they are to rectify. DT&E also provides confidence that the system, as designed, is likely to meet the SyRS requirements. The designers of the system also use DT&E results to validate the design process and minimize the risks associated with the system design.

DT&E covers a broad range of testing levels corresponding to the different design activities conducted during the acquisition phase. At one end of the spectrum, DT&E is responsible for T&E of the lowest level components in the system. As such, a DT&E effort may involve time on the test bench evaluating the initial hardware design mocked up on a breadboard. At the other end of the spectrum, full system prototypes (as described in Chapter 4) undergo T&E as part of the DT&E program prior to the system being offered for AT&E and subsequent transition into the utilization phase. DT&E on system prototypes confirms that integration has been successful and provides confidence that the design meets requirements.

Responsibility for the planning and conduct of DT&E normally lies with the supplier/performance activity. The acquirer organization normally places contractual requirements on the contracting organization with respect to T&E and invariably wants visibility into DT&E activities. Visibility is normally provided during the design reviews, but it is not unusual for the acquirer/tasking activity to be invited to witness DT&E activities and to be provided with copies of any DT&E reports and results. The acquirer may, in fact, consider accepting some DT&E results in lieu of repeating expensive and time-consuming tests procedures during AT&E.

DT&E tapers off as the system nears completion and initial delivery to the acquirer; however, the development of modifications during the utilization phase revives the requirement for DT&E.

5.3.2 AT&E

AT&E, a discrete activity normally shared between the acquirer and supplier, aims to verify that the developed system meets the requirements of the SyRS that drove its development. The acquirer approves the specific acceptance test procedures to be completed as part of AT&E, but both the acquirer and the supplier may jointly conduct the tests. Even if customers do not conduct the AT&E testing, they will most likely witness the process and analyze the results. AT&E, which is focused on the system-level requirements contained in artifacts such as the SyRS and the contract, is designed to verify that the system meets the requirements as contained in the FBL.

The test results and any discrepancies are documented and rectified where required prior to the conclusion of AT&E. Regression testing may be required where some acceptance test procedures have failed. The supplier will perform the necessary modifications to the system, and the test procedures will be repeated. Care must be taken in these circumstances to understand what additional impact system modifications may have on other aspects of system performance. Regression testing should be conducted on all aspects of the system affected by the modification, which may involve significant effort.

At the successful conclusion of AT&E, the acquirer accepts the system from the supplier, and the system formally enters the utilization phase of its life cycle.

5.3.3 OT&E

OT&E is focused on operational functionality of the capability rather than design issues of each constituent element and is therefore commonly the responsibility of the customer organization. OT&E is normally conducted by an entity within the customer organization to assess the ability of the system and its components to meet the BNR/SNR and the StRS. The testing entity is normally independent from the procuring agency within the customer organization to ensure that some level of independent assessment is provided. Separation of OT&E responsibilities from the customer's project office assists in maintaining objectivity in the assessment.

OT&E aims to estimate the operational effectiveness and suitability of the system when it is combined with other capability system elements such as personnel, support, and training and with external systems in the desired operational environments. Potential modifications may be identified during the OT&E activities, and recommendations are passed through the appropriate channels to the designers. OT&E must be conducted in as realistic an environment as possible. Whereas DT&E, and AT&E to a certain extent, can be performed in clinical conditions, OT&E must be conducted in operational conditions including the use of all of the inputs to capability together with operators, support personnel, maintainers, and logistics support.

In this way, OT&E supports true validation of the system, because the original need for the system should be satisfied when the customer's operational personnel operate the equipment that is supported by the customer's support personnel. These operational and support personnel, who will have to be trained via the customer's training system, are doing their jobs in accordance with the customer's processes, procedures, and doctrine.

There is another important distinction between AT&E and OT&E. As illustrated in Figure 5.2, AT&E is conducted at the end of the acquisition phase of each of the elements of the capability system in order to ensure that the contractor has met the requirements for each element of the relevant SyRS for that element; AT&E is therefore principally an act of verification to ensure that the supplier has met its obligations. OT&E is an act of validation that ensures that all the elements of the capability have been integrated and that the delivered capability system meets the stakeholder requirements in the StRS (BNR/SNR).

OT&E may also be used to assist the acquirer in developing operational procedures for the system. Moreover, the system continually undergoes OT&E

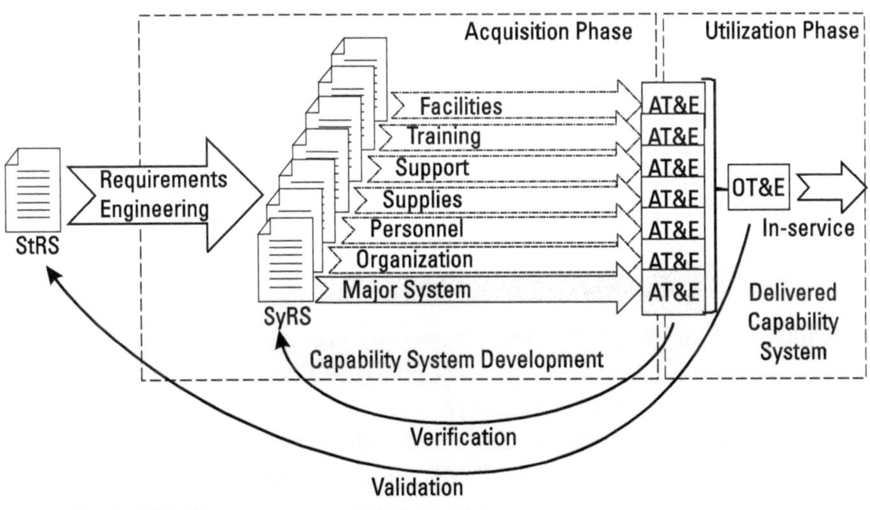

Figure 5.2 AT&E verifies that the system meets the requirements in the SyRS; OT&E validates that the capability system meets the StRS.

throughout its life in service, albeit often informally as operations reveal shortfalls in performance and suggest areas for improvement.

Example 5.1: Testing in Our Aircraft Example

> *In our aircraft example, the company acquiring the aircraft (ACME Air) may be organized into three functional areas—operations, maintenance and support, and procurement. It is normally the responsibility of the procurement section to manage the project associated with the procurement of the new medium-range aircraft. To that end, procurement is responsible for acquirer T&E. Part of their role as the acquirer T&E agent is to involve operations personnel in the T&E effort, particularly in AT&E and OT&E. Through their T&E efforts, the company has much greater confidence as an acquirer for the new aircraft.*

Example 5.2: The Difference Between Verification and Validation

> *During the AT&E process, specialized personnel like test pilots, flight test engineers, and ground-based testing specialists will operate the new aircraft in controlled conditions to execute the various acceptance test procedures and collect relevant test data to support the verification process. When the aircraft has completed AT&E successfully, we can be confident that the aircraft meets all its specified requirements (that is, that the aircraft design has been verified).*
>
> *However, when the aircraft enters operational service with ACME Air, it will be operated by ACME Air pilots and maintained and serviced by ACME Air engineers and technicians in ACME Air maintenance facilities. It will be operated in operational conditions that are not necessarily the same as the controlled environment associated with AT&E. It is only when the aircraft is operated in this way that ACME will be able to assess whether the aircraft is meeting the needs of the new medium-range aircraft capability. The aircraft design will be validated if it is found to meet ACME's needs under these conditions.*

5.3.4 Planning and Managing the T&E Program

The TEMP is the major planning artifact for the entire T&E effort planned for the acquisition phase of the project. The TEMP is also commonly referred to as the Master Test Plan (MTP) or the V&V plan. The TEMP is required by the contract/agreement, prepared by the supplier, and reviewed and approved by the acquirer organization. The TEMP should be finalized and approved early in conceptual design and evaluated as part of the technical review process. The TEMP should be in an approved state by SDR and should be maintained throughout the acquisition phase. In some cases, the acquirer may develop a separate acquirer TEMP to cover all phases of the system life cycle, and the supplier will be required to prepare their own plan that concentrates only on their contractual scope of work.

The content of the supplier's TEMP will be specified by the contract/agreement, which may also mandate a template for the layout. A typical TEMP may be required to contain the following information [8]:

- *System description:* This section includes a summary of the mission description of the system being developed and a description of the system that has been designed to meet the requirements. Major TPMs are listed in prioritized order to reinforce the mission description.
- *Program summary:* The program summary should detail the T&E responsibilities of all organizations participating in the T&E effort. The schedule of the T&E effort should also be outlined in this section of the TEMP to ensure that relevant organizations are aware of the proposed T&E activities.
- *T&E outline:* The TEMP should refer to DT&E, AT&E, and OT&E in this section. Critical technical characteristics (DT&E) and operational issues (OT&E) should be detailed, and any testing conducted to date should be recorded in the TEMP. As well as recording testing to date, the TEMP should detail the future testing (DT&E, AT&E, and OT&E). This way, the TEMP becomes a working artifact for the entire T&E effort.
- *T&E resource summary:* This section details the resources required for the entire T&E effort. Resources include personnel (both acquirer and supplier personnel); specific test and support equipment; test facilities; any simulators, models, or testbeds required; and any necessary personnel training required prior to testing. A schedule should also be included to show when the specific resources are required.
- *Appendixes:* Any other artifacts relevant to the T&E effort should be attached to the TEMP. The set of test procedures that will be used to test the system should either be attached to the TEMP or referenced by the TEMP. Test procedures include the setup procedure and initial conditions and a detailed set of instructions needed to run the test and provide the expected results for each of the tests. The test procedures will also list any specialized facilities or test equipment needed to perform the test, and may also cover the necessary qualifications and experience of test personnel. Test reports are generated following testing and contain a record of the test results, a determination of success or otherwise of the test and recommendations should the system being tested fail the test.

These reports (when completed) will also be either attached or referenced in the TEMP.

The TEMP may also reference other plans and artifacts (such as discrepancy reports) necessary to support the T&E process.

5.4 Technical Risk Management

Risk is defined as the possibility of a loss or injury, or the possibility of some disadvantage or destruction. Systems engineering management is concerned with the management of technical risk or the risk associated with the technical aspects of the system life cycle. Broadly speaking, there are two major categories of risk: internal risk and external risk. Internal risks are those that are within the control of the project office; external risks are beyond the control of the project office. An internal risk could be, for example, the risk associated with inappropriately manning the project office with qualified staff. This risk can be managed from within the project office. An external risk could be the introduction of some new legislation that places more stringent requirements on the system, leading to some changes to the user requirements.

The concept of risk is a subjective assessment in that risk identification and assessment is based on the perception of an individual and their interpretation of the term risk. What one individual may find risky, another individual may not. To that end, risk management must be performed by personnel with a broad and experienced knowledge of the subject area to ensure that risks are assessed realistically. Given the extremely broad nature of risk, it is unlikely that a single person will be able to perform the entire risk management function. Risk management is, therefore, a team effort.

There are several components that combine to define the concept of risk. These risk components (based on PMI's *Project Management Body of Knowledge* [9]) are described as follows:

- *Risk identification:* The first step in the risk management process involves the identification of the potential risks relating to the system. Once the risks have been identified, they can be analyzed and managed accordingly.
- *Risk quantification:* Risk quantification consists of two parts: impact and probability. Risk impact estimates the likely damage a risk would have on the system should the risk occur. Probability relates to the likelihood of the risk occurring. The severity of any particular risk is a function of both impact and probability.

- *Risk response development and control:* The final component of risk management is the management (response development and control) of the risks identified and quantified by the previous steps. Management of the risks involves ranking them and devising a means to deal with the most critical of the risks. Classic means of dealing with risks include avoidance, mitigation, or acceptance.

Technical risk management is not a distinct systems engineering activity assigned to an individual risk manager, but rather an integrated part of a sound systems engineering management effort. This is demonstrated by the fact that many tools and concepts described in this text as systems engineering processes or management functions contribute to technical risk management. Examples include the process of TPMs, T&E, technical reviews and audits, and configuration management. Importantly, technical risk management is not a one-time exercise conducted early in the acquisition phase of the system.; rather it is an ongoing concern. New risks may arise, existing risks may change, and old risks may disappear during a system life cycle. To that end, technical risk management must continue throughout the system life cycle. Most common systems engineering standards and project management standards contain guidance on risk management.

5.4.1 Risk Identification

Risk identification is the process of identifying the major areas of risk within a given system and recording those risks. The process should record the source of risks and the symptoms that may be evidence of an impending risk occurring. Typical sources of risk in projects include changes in customer requirements leading to design changes, design errors, or misunderstandings leading to unsatisfactory performance, and insufficiently skilled staff leading to longer design times or substandard designs.

No foolproof set of tools exists that guarantees the complete identification and recording of all risks. Potentially the most effective means of risk identification is via a series of interviews or meetings with the appropriate groups of people. Given that both internal and external risks exist, the skill is in identifying the appropriate groups of subject matter experts to interview.

Any source of information may provide insight into potential problems and risks. Sources of information include systems engineering design artifacts such as drawings and system architectures, trade-off studies, or design review artifacts. The other risk management tools such as audit results, TPM monitoring, and T&E results also provide invaluable insight into potential new risks and the performance of existing risks.

It may be helpful to consider the following major categories as potential sources of risk during the identification process:

- Design, including design process, requirements, analysis, and testing;
- Testing, including the test reports containing test results;
- Production, including the manufacturing plan and process, test equipment requirements, and subcontractor controls;
- Logistics, including facility requirements, manpower and personnel, spares, and technical documentation.
- Management, including cost and schedule estimates, acquisition strategies, and staffing considerations.

By using these major categories, it may be simpler to assign responsibility for managing the identified risks.

Technical reviews and audits are also an excellent forum for discussing risk and potentially identifying new risks as the system moves through the acquisition phase. Strong consideration should be given to contractually requiring risk management activities in each of the major formal reviews and audits.

5.4.2 Risk Quantification

Once the major risks have been identified, an analysis of each risk needs to be carried out to determine the likelihood of the risk occurring and the impact of that risk should it occur. This is called risk quantification. Based on the assessments of likelihood and impact, a severity of the risk (which is a function of both likelihood and impact) can be calculated. By attempting to analyze aspects such as likelihood and impact, the quantification exercise will focus on the cause and effect of the risk and in doing so provide some insight into the potential management of the risk.

Several tools exist to assist in the analysis of risks including classic project management scheduling tools such as program and evaluation review technique (PERT) and critical path method (CPM) charts, simulation, and models. There is benefit in using available tools to provide an input into the risk quantification, but expert judgment remains a key ingredient in risk analysis. It is vitally important to maintain focused on the fact that risk quantification is primarily subjective in nature. There is a lot of merit in keeping the risk quantification relatively simple. Its main aim is, after all, to provide management with a list of potential risks (in order of severity) and provide some indication of the causes and effects of these risks. Management will then decide on the best available means of managing these risks as discussed in Section 5.4.3.

A simple but effective means of ranking the severity of the risks is to categorize the likelihood of occurrence into three categories: highly likely, possible, or unlikely. Using the same process, the impact of the risk should also be categorized into three categories; high, moderate, or low. In this way, there will be nine possible levels of risk severity. It is common to use five categories leading to 25 levels of severity. To make the risk assessment practical, it may be advisable to use less than nine levels of severity. Excessive levels add confusion without greatly adding to the ranking exercise. For example, it may be decided to use only three levels (high, medium, and low) as shown in Figure 5.3. By limiting the number of severity levels to three, the relative ranking of the risks is simplified.

5.4.3 Risk Response Development and Control

There are three main options for the management of the risks identified and quantified in the previous two steps: risk avoidance, risk mitigation, and risk acceptance. The management of risk should not focus solely on the negative aspects of risk, but also on the benefits that may accrue from exposing the project to the risk.

Risk avoidance is the option selected when the risk is too severe to be accepted at any level and must be avoided. Risks that may fall into this category are those that may result in loss of life or loss of business. The causes of the risk will need to be investigated to decide on how best to avoid the risk. It should be noted that by avoiding one risk, exposure to other risks may well result. This additional exposure also must be considered when deciding on how best to avoid a risk.

Risk mitigation may be the option selected when the risk falls in the middle ground in terms of severity. Mitigation basically means weakening, and, in the context of risk, mitigation means reducing severity. This can be achieved by reducing the impact of the risk, reducing the likelihood of its occurrence, or

		Likelihood		
		Highly likely	Possible	Unlikely
Impact	High	High	High	Medium
	Moderate	High	Medium	Low
	Low	Medium	Low	Low

Figure 5.3 Risk severity table.

preferably both. Again, the causes and effects of the risk must be considered to mitigate the risk successfully.

Risk acceptance is also a valid form of risk management usually reserved for risks of low severity. Risks may also be accepted when there is significant benefit to be gained should the risk *not* occur. The adage of risk-equals-return does hold true sometimes for technical risks. For example, the use of leading-edge technology may be judged to be a reasonably severe risk as the impact of it going wrong may be quite high. However, management may decide to go ahead with the use of the technology, because, if it does not go wrong, the benefits may be extremely attractive.

Risk response development and control also considers how to monitor the risks identified. Severe risks should be monitored more closely and more often than less severe risks. However risks are managed, they must be managed continually to address the fact that circumstances are continually changing; old risks will disappear, new risks will appear, and existing risks will change in nature. Risk management, the responsibility of all parties involved in the system life cycle, must be performed continually.

5.4.4 Risk Management Records

If there appears to be significant risk associated with a given system development, there may be a requirement for a risk management plan (RMP), which is a formal record written by the contracting organization and reviewed and approved by the acquirer organization. The RMP details the risk management process and the results of the application of that process for the development effort. In addition, the acquirer organization develops a more general RMP associated with all project management–related aspects of system development.

The risk management process should, as a minimum, contain the major steps outlined in this section, and the results (of the identification and quantification) will depend upon the nature of the system development. The RMP should detail the schedule for reviewing the risks and should contain details of the management (response and control) of each of the risks. The plan must be periodically reviewed, since it is likely that risks will change with time.

5.5 Configuration Management

The term "configuration" can be defined as the relative disposition or arrangement of parts of something. In the context of systems engineering, the "something" is normally a CI that forms a part of the overall system, and the "relative disposition or arrangement of parts" is called a baseline. To that end, configuration management (CM) is the act of controlling and managing the physical and functional makeup (defined in the baselines) of the configuration items that comprise the system [10]. EIA-649-C defines configuration management as, "A technical and management process applying appropriate processes, resources, and controls, to establish and maintain consistency between product configuration information, and the product" [11].

The main aims of CM are [12] to identify the functional and physical characteristics of selected system components designated as configuration items; to control changes to those characteristics; and to record and report on the change processing and implementation status.

An effective CM system provides an accurate snapshot of the state of each CI (and through these, the system). CM allows an effective history of changes to the system design to be maintained, including the alternatives considered and the reason for selecting the preferred alternative.

5.5.1 Establishing the Baselines

The baselines used in the CM effort, discussed in Chapters 2 to 4, are a very important part of system acquisition, as each approved baseline provides a firm position from which to move forward. The three baselines involved in CM are described as follows.

- *FBL:* The FBL of the system is defined principally by the approved SyRS produced during the early stages of the acquisition phase. Once approved, the FBL will be placed under configuration control, which means that changes to the baseline can only occur through a formal change management process. This process is different for each project, but an example change management process is described in this section. The FBL is eventually finalized when all FCAs have been completed prior to the utilization phase.
- *ABL:* The ABL for the system comprises the set of development specifications and ICDs (if applicable) for each of the CIs making up the system. These specifications should be developed and approved during the preliminary or detailed design activities of the acquisition phase. The ABL is initially established when the set of development specifications is complete, and finalized or approved for each CI when the relevant FCA has been conducted successfully.
- *PBL:* The PBL for the system includes the set of product specifications (and any associated material and process specifications) for each of the CIs comprising the system. The stage at which this baseline is established can vary but is typically towards the end of the detailed design and development stage or very soon thereafter. The PBL is then confirmed or approved by at the conclusion of PCA.

5.5.2 CM Functions

For CM to be effective, the following functions must be performed throughout the project:

- CM planning and management whereby the CM process is planned, recorded, and approved [normally in a governing plan called the configuration management plan (CMP)];
- Configuration identification of the items to be placed under CM;
- Configuration change management of these items through an effective and formal change management process;
- Configuration status accounting to provide accurate and up-to-date status information on the configuration of the items placed under control;
- Configuration verification and audits (such as PCA and FCA) to confirm the correct operation of the CM system.

5.5.2.1 Identifying the CIs

Configuration identification revolves around the selection of the CIs during the early activities of the acquisition phase. Selection of CIs is a design decision recorded through the breakdown of the SyRS into lower level specifications such as product and design specifications. Typically, systems are broken into the major categories of computer software CIs (CSCIs) and hardware CIs (HWCIs). Naturally, there may be several CSCIs and HWCIs in the system design, but each CI is managed as a separate entity including the design, documentation, and T&E of the item.

The selection of CIs is based on a number of factors, described as follows:

- *Complexity:* One of the aims of the systems engineering process is to reduce the design problem to be manageable. Large complex CIs defeat this purpose, so CIs should be small enough to be able to be designed and developed by an individual or a small team.
- *Interfaces:* The difficulty with creating small, interconnected CIs is that the larger number will add additional interface definition and management. The selection of CIs is therefore a trade-off between the complexity of each CI and the number of interfaces required.
- *Use/function:* Each CI should perform a single function or a combination of related functions.
- *Existence of available items:* Rarely will a system design require all components to be completely designed and developed. Many of the components required by a system will already exist in a certain configuration,

and the design should ensure that such components are used rather than unnecessarily reinventing products.

- *Supplier:* Similarly, unnecessary complexity is introduced into the design if a single CI is provided by more than one supplier.
- *Criticality:* A CI may be chosen to be smaller than would otherwise be required if the designers feel that the item is critical and therefore needs to be managed separately.
- *Maintenance and documentation needs:* Through-life support is an essential consideration in design and CIs should be chosen to facilitate maintenance and documentation.

5.5.2.2 Managing Configuration Changes

At various stages throughout the system development, changes may be required. The reasons for these adjustments will be wide and varied, ranging from a modification to one of the requirements to an alteration in some available technology. Changes must be managed by a formal process to ensure that the impacts of the change are investigated and handled correctly.

Normally, a body called the configuration (change) control board (CCB) has responsibility for managing changes. The CCB investigates the change and decides on the severity of the change. Major changes (sometimes called class I changes), which involve shifts in form, fit, or function of the system, or major adjustments to system cost and schedule, generally involve a robust change approval process. Minor changes (class II changes), which include all other modifications to the system configuration, generally go through an abridged change management process. Examples include updates to documentation following identification of errors or changes in aspects of the design that do not affect form, fit, or function, such as material substitutions. The CCB meets on a regular basis to review the progress of changes that are currently in progress and to review other proposed changes but can meet in between regular meetings to process a change that is considered urgent.

A change proposal must contain sufficient information for the CCB to decide whether to approve the change. The change proposal includes a statement outlining the change and the reasons that the change has been suggested. The proposal then details the alternatives available to implement the change and outlines the preferred alternative. The preferred alternative should be provided in sufficient detail to show that it will result in the change being implemented successfully and not introduce new problems into the system design. Of particular importance is the impact of the change on other CIs within the design. The effect of the change may result in changes to the interface requirements as specified in the baseline documentation. To that end, changes to the baseline documentation should also be included in the change proposal. The ability to assess impact relies heavily on sound traceability across the different levels of design.

A simple change management process is shown in Figure 5.4.

In some cases, it becomes clear during construction or production that items within the supplies are not going to comply with the relevant specification contained in the established configuration baseline. There may be any number of reasons for this to happen. Normally, these nonconformances would be rectified using the formal engineering change process; however, there are situations when departures from specifications may be accepted by the acquirer without the need to go through that process. CM calls these special cases requests for variance [13] (previously called deviations or waivers [14]) or concessions [15]. Note that a variance differs from an engineering change in that an approved engineering change requires corresponding revision of the CI's current approved configuration, whereas a variance does not.

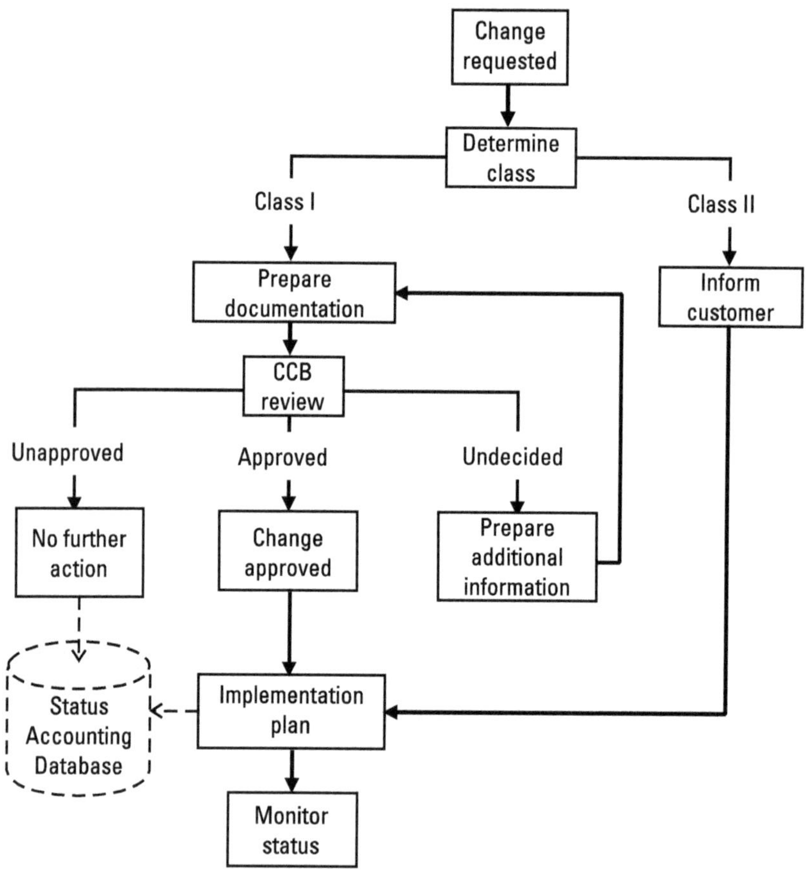

Figure 5.4 A simple change management process.

Example 5.3: Request for Variance

Following the analysis of requirements, a design is developed that includes a particular OTS computer system. When the time comes to procure and integrate the OTS computer, it is not available in time to support continuing integration, testing, and initial acceptance of the system. An alternative OTS computer system of similar specifications may be available. A request for variance may be sought and approved prior to the procurement of the alternative computer system to allow the designers to use that computer system until the specified computer is available. In this example, the acquirer would be accepting an alternative for a specified period. There is no need to seek an engineering change because the contractors are not seeking to change the configuration baseline of the system, nor is the change permanent. Once the preferred OTS computer becomes available, it will be integrated in accordance with the approved baseline.

5.5.2.3 Other CM Functions

Status accounting is the function that provides information about the status of the items identified as requiring configuration control. This information includes the current state of the CIs and a history of any changes to the CIs from initial identification. Effective status accounting significantly supports the change management process, which in turn allows changes to be made to the various system baselines during the design process.

Configuration audits (discussed in Section 5.2), including FCA and PCA, confirm that the CM process is operating effectively. FCA confirms that the CI possesses all necessary functionality, whereas PCA concentrates on the physical configuration of the CI as described by the relevant documentation and displayed by the physical CI. For items of software, the final versions of software listings are normally audited to ensure that the executable code is produced by a compilation of the documented source code.

5.6 Specifications

One of the major outputs from the systems engineering processes is a complete set of design and technical data that accurately describes the designed system. This technical information and data is normally contained in design artifacts called specifications. Specifications are produced (or procured) during the acquisition process to describe the system or its components to varying levels of detail. Note that a specification can be a document (either physical or electronic), a structured database produced by one of the many commercially available requirements management tools, or a collection of other data sources, such as models produced using MBSE tools and processes. Specifications are generally structured around the system hierarchy and often illustrated as a specification tree. An example is shown in Figure 5.5 with the specifications described in Chapters 2 to 4. In this case, the specification tree shows how lower level specifications flow from, and are related to, the SyRS.

Project managers need to be aware that every system development project will have some requirements associated with generating technical information and data. Significant effort and resources will be required to generate, review, and maintain this information and data, and this effort needs to be included in the project scope and estimated in the cost and schedule provisions in the WBS and the master schedule.

5.7 Standards

In addition to technical information and data relating specifically to the system being designed and produced, published industry standards may be applied to the design of complex systems. In some cases, standards will be used, because they represent best practice, but in other cases, they may be contractually man-

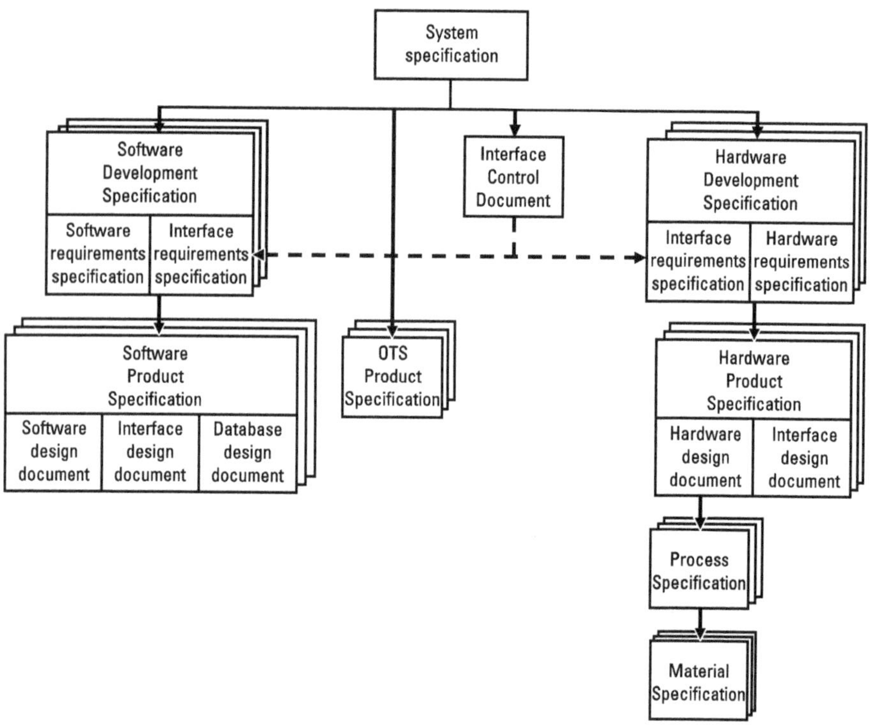

Figure 5.5 Typical specification tree.

dated. These standards take many forms including military standards, international standards, and standards from recognized institutions such as the Institute of Electrical and Electronic Engineers (IEEE), the International Standards Organization (ISO), and the Electronic Industries Alliance (EIA). Some classic examples of standards include the following:

- Safety standards (for example, ISO 45001);
- Environmental standards (for example, ISO14001 and MIL-STD-810);
- Reliability standards (for example, MIL-HDBK-217);
- Technical interface standards (for example, IEEE 802.3 and RS-485);
- Human factors standards (for example, MIL-STD-1472);
- Design standards (for example, ISO/IEC/IEEE 12207):
- Management standards (for example, ISO 21500, ISO/IEC/IEEE 15288, ISO 9001, and ISO 45001).

While these kinds of standards bring together a wealth of knowledge and experience and provide system designers with technical guidance during system development, they can also add significantly to the development effort.

When considering the use of standards, there are some risks to be aware of. For example, there is a risk of overspecifying engineering standards in the belief that including more stringent requirements somehow enhances the chances of success of the system development. Specifying engineering standards that are not truly required introduces significant overhead without necessarily reducing any risks associated with the development. The increased overhead is passed directly onto the acquirer in the form of extended schedules and increased costs. To overcome this risk, standards specified in contractual documentation should be there for a reason and not merely there because "that standard has always been in our contracts."

A risk related to overspecification is the "blanket specification" of engineering standards. This means specifying the entire standard without consideration of the relevance (or lack thereof) of each of the sections within that standard. Standards are often designed to be tailored to meet the specific requirements of each development project. For example, certain technical reviews and audits may not be necessary in a particular project and should be explicitly "tailored out" of the standard that is being used to specify the review and audit requirements of the project. Without tailoring, unnecessary (and therefore expensive) activities and requirements may be undertaken by the design and development team. Effective tailoring can only be done by individuals who are conversant with the standard, the overall systems engineering effort, and the specific system in question.

A final risk to be aware of is that standards sometimes reference other standards where necessary rather than repeating information that has been documented elsewhere. For example, standard "A" may make reference to the requirements in standards "B," "C," and "D." Standard "B" may go on to reference the requirements in standards "E," "F," and "G," and so on. Therefore, by requiring the use of standard "A," the acquirer organization may unintentionally invoke the requirements of standards "B" through to "G." This is another reason to carefully select and tailor any standards required by the contract.

5.8 Integration Management

In large and complex system developments, a wide range of technical specialists (or experts) will be involved with producing a wide variety of subsystems. The specialists perform work in the traditional engineering fields such as electrical, civil, mechanical, and aeronautical engineering, and in very specialized and detailed areas such as EMI/EMC and corrosion control. With all these specialized areas working on the same project, there is a significant requirement for a coordinated approach to the integration effort. The integration effort needs to be managed to ensure that compatibility exists between the different areas. The task of systems integration falls within the bounds of systems engineering management and can be managed effectively by ensuring that adequate attention is paid to the functional and physical interfaces between the different design areas.

The task of integrating different system elements is managed primarily through an artifact called the ICD, which was introduced in Chapter 3. The ICD is responsible for specifying the functional and physical interface requirements that exist between the different subsystem elements within the system. The detailed requirements for those interfaces are then defined in both the development and product specifications for each of the respective CIs. The ICD, however, identifies the CIs between which there needs to be an interface, and through the definition of that interface, flags the need for communication to occur between the different design teams. This is a critical process, because interfaces often result in additional derived requirements being placed on the subsystems. Formal interface control is a prerequisite for successful integration.

Example 5.4: Importance of the ICD

Suppose the system consisted of CI A and CI B. There would be separate development and product specifications for both A and B. The team responsible for A would not necessarily access the development and product specifications for B and vice versa. A single ICD would be produced (called the A-B ICD or similar) defining the common interface requirements. Both teams would need to refer to this ICD because it will refer to both products. Both teams would also be responsible for cooperatively producing and maintaining the A-B ICD. The ICD becomes the conduit through which the A and B teams communicate and therefore assists in the systems integration effort.

The interfaces are defined in gradually increasing levels of detail as the design progresses from conceptual design, through preliminary design and onto detailed design and development. Example 5.4 illustrates the fact that design teams from different interfacing areas (that is software, hardware, and specialty engineering areas) need to work as a team to bring the interface definition and design together. These teams may be called ICWGs, although other terms exist. ICWGs must meet regularly to generate the initial ICDs and should then meet periodically or as the need arises to address any ongoing issues. These meetings and reviews may be informal, but the products of the meetings are formal interface design issues and decisions, and updated ICDs.

ICDs and minutes from ICWG meetings may be formally reviewed during the formal technical reviews and audits process outlined in Section 5.2.

5.9 Systems Engineering Management Planning

Systems engineering covers so many technical and management topics that a governing SEMP, often called an engineering plan, is generally formulated to plan and document the approach. The SEMP is normally proposed by the design and development organization in accordance with the requirements in the contractual documentation and is then reviewed and approved by the acquirer. Once approved, the SEMP becomes the authoritative plan controlling the entire systems engineering effort. Changes to the SEMP must be reviewed and approved as they occur, and the SEMP is normally reviewed at each of the formal design reviews.

The SEMP is the fundamental artifact when evaluating the ability of a supplier to perform the technical activities associated with the system development project. Suggested contents for the SEMP are provided by several systems engineering standards [16] reviewed in Chapter 6.

The SEMP should cover all the major systems engineering functions. It may do so by referring to other subordinate plans such as the CMP, the RMP, the TRAP, and the TEMP; specification lists/trees; CI lists; and so on. Internal company design plans and processes may also be referenced by the SEMP if applicable to save on duplication and repetition. In this way, the SEMP helps to define the engineering management and processes to be applied to the project. In addition to engineering management and processes, the SEMP should also detail positions of responsibility within the design team, including chief designers, software and hardware team leaders, project managers, testing personnel, and so on. Part of the SEMP approval process should include an assessment of the skills and qualifications of these key personnel. Naturally, if this information is found in another management plan, the SEMP may refer to that plan.

The content of the SEMP should be maintained throughout the system design and development effort. Changes to the SEMP must be approved by the acquirer, as this ensures visibility into changes that may expose the project to unexpected risks (such as a change in key personnel).

5.10 Summary

This chapter considers the major systems engineering management elements of technical review and audit management, T&E, technical risk management, CM, the use of specifications and standards, and systems engineering management planning. Systems engineering management is vitally important, but care must be taken to tailor the systems engineering management activities to suit the project, else effort will be wasted or, worse, additional risk will be created by overmanagement.

Review Questions

1. Explain why technical reviews and audits are vital to systems engineering management and describe the major formal reviews and the audits that may occur prior to construction and/or production.
2. List and describe the three categories of T&E and describe the primary focus of each of them.
3. Briefly describe the major steps in the risk-management process.
4. List and explain the four main options for the treatment of risk, and explain when each option may be preferred over the others.
5. CM involves five major functions. List these functions and briefly describe them.
6. Explain the role of the CCB in the context of configuration management.
7. Briefly describe the role and content of the SEMP.

Endnotes

[1] Defense Systems Management College, *Systems Engineering Management Guide,* Washington, DC: US Government Printing Office, 1990, p. 12-3.

[2] MIL-STD-499B, *Military Standard—Systems Engineering—Draft,* Washington, DC: US Department of Defense, 1994.

[3] MIL-STD-1521B, *Technical Reviews and Audits for Systems, Equipment, and Computer Software,* Washington, DC: US Department of Defense, 1985.

[4] Examples of commercially available standards and guides that contain information on technical reviews and audits include the following.

ISO/IEC/IEEE 15288, *Systems Engineering—System Lifecycle Processes,* International Standards Organization, 2023.

ISO/IEC/IEEE 15288.2, Application of Technical Reviews and Audits for Defense Programs, 2023.

Walden, D. D, et al. (eds)., *Systems Engineering Handbook: A Guide for System Lifecycle Processes and Activities,* INCOSE-TP-2003-002-05, Hoboken, NJ: John Wiley & Sons, 2023.

[5] MIL-STD-973, *Military Standard—Configuration Management,* Washington, DC: US Department of Defense, 1992.

[6] MIL-HDBK-61B, *Military Handbook—Configuration Management Guidance,* Washington, DC: US Department of Defense, 2020.

[7] EIA-649C, *Configuration Management Standard,* Warrendale, PA: SAE International, 2019.

[8] Defense Systems Management College, *Systems Engineering Management Guide,* Washington, DC: US Government Printing Office, 1990.

[9] Project Management Institute Standards Committee, *A Guide to the Project Management Body of Knowledge* (Version 6), Upper Darby, PA: Project Management Institute, 2017.

[10] CM is a well-described process. Some useful sources of detailed information on traditional and modern CM practices can be found in the following:

EIA-649C, *Configuration Management Standard,* Warrendale, PA: SAE International, 2019.

EIA-836, *Configuration Management Data Exchange and Interoperability.*

GEIA-HB-649, *Implementation Guide for Configuration Management.*

MIL-HDBK-61B, *Military Handbook—Configuration Management Guidance,* Washington, DC: US Department of Defense, 2020.

MIL-STD-973, *Military Standard—Configuration Management,* Washington, DC: US Department of Defense, 1992.

ISO 10007, *Quality Management Systems—Guidelines for Configuration Management,* 2017.

[11] EIA-649C, *Configuration Management Standard,* Warrendale, PA: SAE International, 2019.

[12] Defense Systems Management College, *Systems Engineering Management Guide,* Washington, DC: US Government Printing Office, 1990, p. 11-1.

[13] EIA-649C, *Configuration Management Standard,* Warrendale, PA: SAE International, 2019.

[14] Other sources of information on deviations and waivers include the following:

MIL-STD-973, *Military Standard—Configuration Management,* Washington, DC: US Department of Defense, 1992.

MIL-HDBK-61A(SE), *Military Handbook—Configuration Management Guidance,* Washington, DC: US Department of Defense, 2001.

MIL-STD-481B, *Military Standard Configuration Control—Engineering Changes (Short Form), Deviations and Waivers,* Washington, DC: US Department of Defense, 15 July 1988.

[15] ISO 10007, *Quality Management Systems—Guidelines for Configuration Management,* ISO, 2017.

[16] There are a number of standards that suggest contents of a SEMP, including the following:

MIL-STD-499B, *Military Standard—Systems Engineering—Draft,* Washington, DC: US Department of Defense, 1994.

DI-S-3618/S-152, Data Item Description, *Systems Engineering Management Plan (SEMP),* Washington, DC: US Department of Defense, February 9, 1970.

EIA/IS-632, *Systems Engineering,* Washington, DC: Electronic Industries Association (EIA), 2003.

EIA-632, *Processes for Engineering a System,* Washington, DC: Electronic Industries Association (EIA), 2003.

IEEE-STD-1220, *IEEE Standard for Application and Management of the Systems Engineering Process,* New York: IEEE Computer Society, 2005.

ISO/IEC/IEEE 15288, *Systems Engineering—System Lifecycle Processes,* International Standards Organization ISO/IEC/IEEE, 2023.

ISO/IEC/IEEE 24748-4, *Systems and Software Engineering—Lifecycle Management—Part 4: Systems Engineering Planning,* 2018.

SAE1001, *Integrated Project Processes for Engineering a System,* SAE International, 2018.

6

Systems Engineering Tools

Systems engineering tools are available to assist both systems engineering management and the systems engineering processes, as well as related disciplines such as project management. This chapter discusses the use of systems engineering standards and INCOSE guidance as management tools and then look briefly at systems engineering process tools. We also briefly consider MBSE.

6.1 Systems Engineering Management Tools

This section outlines several standards that assist the systems engineering management effort. We also summarize the major guidance available from INCOSE.

While there has been considerable movement toward a harmonious suite of standards relating to systems engineering over the last decade, there is some way to go, and there are still several standards guiding system development in organizations around the world. Although ISO/IEC/IEEE 15288 and its attendant handbooks and guides are being rapidly adopted, there are many projects still making use of older standards such as EIA 632 and IEEE-STD-1220. There are also many organizations whose internal processes have been developed based on an older standard, so there is considerable inertia to rapid change.

Consequently, we briefly cover all the major systems engineering standards in this chapter for two reasons. First, they are all in use to various degrees in projects around the world, so the reader will most likely be interested in their use of a heritage standard as well as in the content of the modern standards, handbooks, and guides. Second, we want to illustrate that, while there are clear differences, all the major standards provide a framework for the establishment of useful systems engineering processes and procedures for a project.

Additionally, any selected standard will almost always have a scope greater than the project and system under consideration—so tailoring is essential, not only because of the scope of the standard but also because of the need to temper the selected standard's requirements to accommodate the culture and structure of the adopting organization.

6.1.1 Evolution of Systems Engineering Standards

The evolution of systems engineering standards is shown in Figure 6.1 to illustrate the history and previous standards upon which current standards are based. Sections 6.1.1 to 6.1.9 briefly describe the standards and discuss their evolution as shown in Figure 6.1 from MIL-STD-499B to the current standard ISO/IEC/IEEE 15288, which was most recently revised in 2023. In the following sections, we discuss MIL-STD-499B, EIA/IS 632, IEEE-STD-1220, EIA 632, and SAE 1001.

6.1.2 MIL-STD-499B—Systems Engineering (Draft)

MIL-STD-499B was developed by the US Department of Defense (DoD) and released in draft form in 1994 to replace MIL-STD-499A, which was dated 1 May 1974. The Bravo version of 499 was never released, except in draft form. In 1994 the US government directed that most US military standards were to be canceled if a commercial standard was available. EIA/IS 632 was in development, so it was adjusted to accommodate military requirements, and it can be considered as the demilitarized version of MIL-STD-499B. Even though MIL-STD-499B was not officially released, it was widely considered to be the premier systems engineering standard, and it remains the source of much of the systems engineering language and methodologies still in vogue today.

The SEMP is seen as a central systems engineering artifact as an effective way to tailor the requirements in MIL-STD-499B to meet the specific requirements of the individual project. Both the acquirer (referred to as the tasking activity) and developer (referred to as the performing activity) should produce SEMPs to describe and integrate their individual systems engineering efforts. MIL-STD-499B also requires schedule information to be contained in the SEMP and calls for two scheduling artifacts: the Systems Engineering Master Schedule (SEMS) and the Systems Engineering Detailed Schedule (SEDS). Once approved, the SEMP becomes the key systems engineering artifact for the project.

Although MIL-STD-499B is out-of-date and was never issued, the standard is an excellent source of information and guidance on systems engineering management and technical planning. The contents of MIL-STD-499B reflect a vast and varied range of experience in the field and provide both acquirer and developer organizations with learned advice on engineering management.

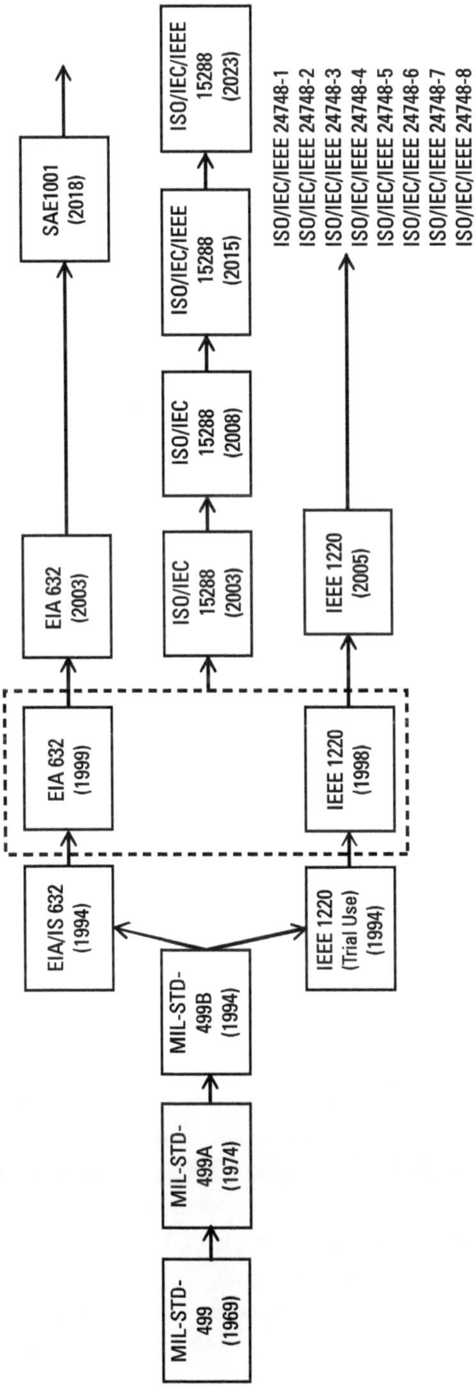

Figure 6.1 Evolution of systems engineering standards.

Consequently, MIL-STD-499B is still a useful source of systems engineering information. For defense projects, MIL-STD-499B has been replaced by ISO/IEC/IEEE 15288.1 and ISO/IEC/IEEE 15288.2.

6.1.3 EIA/IS-632—Systems Engineering

When it became evident that MIL-STD-499B was not going to be released, an established EIA committee undertook the adaptation of MIL-STD-499B for use as an industry standard. The adapted standard, called the EIA Interim Standard (IS) 632, was released in December 1994. Work on the updated standard was completed in 1999 with the release of EIA-632.

EIA/IS-632 is almost identical to MIL-STD-499B. The differences in general content are minor, as is immediately obvious by scanning the respective tables of content. Some military-specific phrases and examples have been altered to make the EIA/IS-632 standard less military-focused. The other changes are very minor and do not make EIA/IS-632 substantially different from MIL-STD-499B.

As with MIL-STD-499B, EIA/IS-632 contains excellent information, advice, and guidance. It is worth remembering that MIL-STD-499B was only ever a draft and was never officially released for use by the US Department of Defense. The major advantage, therefore, of EIA/IS-632 over MIL-STD-499B is that EIA/IS-632 was officially released by the EIA. For this reason, the use of EIA/IS-632 over MIL-STD-499B was recommended. As with MIL-STD-499B, for defense projects, EIA/IS-632 has been replaced by ISO/IEC/IEEE 15288.1 and ISO/IEC/IEEE 15288.2.

6.1.4 IEEE-STD-1220—IEEE Standard for Application and Management of the Systems Engineering Process

The IEEE started work on IEEE-STD-1220 in 1989. Early work on the standard focused on parallel effort with the development of MIL-STD-499B to ensure consistency between industry and military/government systems engineering direction. When released in 1994, IEEE-STD-1220 was the first commercial systems engineering standard so was released as a "Trial Use Standard." IEEE updated the trial use standard into a full standard in December 1998 but retained the designation IEEE-STD-1220. There are only minor differences between the 1994 and 1998 IEEE-STD-1220 standards.

IEEE-STD-1220 has the same scope as EIA/IS-632 in that it is designed to be applied to the technical effort of a wide range of projects covering development, modification, or rectification of equipment. It is applicable to large or small projects, using new or existing technology, producing single or multiple units, being dominated by either hardware or software. IEEE-STD-1220 was intended to be applicable to commercial, government, or military projects.

IEEE-STD-1220 is focused on a system paradigm in that everything within the system (including products, subsystems, and processes) can be dealt with as a smaller system within the main system. This is possible because products, subsystems, and processes all have boundaries, relationships, and interfaces with the other products, subsystems, and processes within the main system. In this way, the requirements of IEEE-STD-1220 can be applied to each individual system within the main system, ensuring a consistent approach to the technical effort behind the project.

IEEE-STD-1220 presents a very familiar systems engineering process that is consistent with the processes adopted by EIA/IS-632 and detailed in this book. The process is applied at each stage of the system life cycle and at each level of system decomposition.

The systems engineering planning requirements contained in IEEE-STD-1220 center around the engineering plan (the equivalent of the SEMP), which is the result of the tailored application of the standard, the master schedule, the detailed schedule, and various technical plans. IEEE-STD-1220 explains that typical plans produced to support the engineering plan include risk management, technical reviews, CM, verifications, computer resources, manufacturing, maintenance, training, security, and human systems engineering. Tailoring is again emphasized as an important part of the application of the standard to different situations and projects.

The importance of design data is emphasized in IEEE-STD-1220 through the requirement for an integrated database to capture and track design data. An integrated data package is also required to record the architecture and design information, as necessary to support manufacturing and through-life support. IEEE-STD-1220 lists and briefly describes the specifications and drawings that are expected in a typical integrated data package. IEEE-STD-1220 refers to the SBS, a tool that takes on the same role as the WBS in the other standards.

IEEE-STD-1220 is a detailed and complete systems engineering standard that provides some excellent systems engineering advice and guidance. Unlike EIA/IS-632, IEEE-STD-1220 is not simply a close replica of MIL-STD-499B but rather a comprehensive coverage of systems engineering. It provides details of the system life cycle and how the different phases are covered by systems engineering and presents a slightly more involved systems engineering process than has been considered in this text. IEEE-STD-1220 is in wide use today but has been replaced by ISO/IEC/IEEE 24748-4, which replaced IEEE-STD-1220 as well as ISO/IEC 26702.

6.1.5 EIA-632—Processes for Engineering a System

EIA-632 [1] was developed jointly by the EIA and INCOSE, approved to be an ANSI standard and released in 1999. EIA-632 was in development since the release of the interim standard EIA/IS-632 in December 1994. Martin [2] provides an excellent overview of EIA-632 and its heritage. The standard represents a significant change from the interim standard and focuses on the enterprise and its policies and procedures rather than focusing narrowly on a particular system development being undertaken by that enterprise. This broad focus sets EIA-632 apart from previous systems engineering standards. In fact, so great is the change that people may not consider EIA-632 to be a systems engineering standard at all—it certainly does not define systems engineering or even use the term.

EIA-632 introduces 13 processes required to successfully engineer a system. Within these 13 processes, there are 33 requirements—see Table 6.1 [3]. EIA-632 focuses on process rather than method. The methods for achieving the requirements are expected to come from enterprise policy and procedures. Further, not all the 33 requirements will be relevant or required in every situation; therefore, EIA-632 is expected to be tailored to individual development projects.

The standard has been designed to help enterprises develop policies and procedures relevant to their role in engineering systems within their industries. Projects being conducted by the enterprise are then responsible for further development of company policies and procedures into project-specific plans, schedules, and so on. By operating at the enterprise level, EIA-632 avoids being a prescriptive standard and concentrates on what must be done to successfully and completely engineer a system. The organization and approach of EIA-632 provides a very detailed framework upon which systems engineering enterprises can base their company policies and procedures.

EIA-632 defines a system as comprising hardware, software, people, facilities, data, materials, services, and techniques. In addition to this definition, a system is proposed to consist of the end product that is the classic "deliverable" but also the so-called enabling products that allow the end products to be produced.

A major concept introduced in EIA-632 is that of the building block. This shows that each end product (of which there may be many per system) can be broken into subsystems. Each subsystem can be considered a system in its own right and dealt with in accordance with EIA-632. As discussed in Section 1.7.1, this forms the basis of the top-down development concept, which breaks the end products into increasing levels of details from layer 1 (system level) down to as many "layers" as required by the development process.

EIA-632 then goes on to describe the bottom-up realization concept, which details how the very low-level layers are progressively integrated and tested until the system-level end product is produced. The top-down development and bottom-up realization concepts are not unique to EIA-632 and form the

Table 6.1
The 33 Requirements from EIA-632

Supply Process Requirements	**Requirements Definition Process Requirements**	**Systems Analysis Process Requirements**
1. Product supply	14. Acquirer requirements 15. Other stakeholder requirements 16. System technical requirements	22. Effectiveness analysis 23. Trade-off analysis 24. Risk analysis
Acquisition Process Requirements	**Solution Definition Process Requirements**	**Requirements Validation Process Requirements**
2. Product acquisition 3. Supplier performance	17. Logical solution representations 18. Physical solution representations 19. Specified requirements	25. Requirement statements validation 26. Acquirer requirements validation 27. Other stakeholder requirements validation 28. System technical requirements validation 29. Logical solution representations validation
Planning Process Requirements	**Implementation Process Requirements**	**System Verification Process Requirements**
4. Process implementation strategy 5. Technical effort definition 6. Schedule and organization 7. Technical plans 8. Work directives	20. Implementation	30. Design solution verification 31. End product verification 32. Enabling product readiness
Assessment Process Requirements	**Transition to Use Process Requirements**	**End Products Validation Process Requirements**
9. Progress against plans and schedules 10. Progress against requirements 11. Technical reviews	21. Transition to use	33. End products validation
Control Process Requirements		
12. Outcomes management 13. Information dissemination		

basis for most popular systems engineering thinking on system development (as described in Chapter 1).

EIA-632 provides several useful annexes. Annex A is a standard inclusion listing the glossary of terms used throughout. Annex B provides additional detail on the engineering life cycle used throughout the standard. Annex C provides a very useful description of each of the requirements in terms of representative tasks needed to achieve the requirements and the expected outcomes from each of the representative tasks. Annex D provides a description of the documentation that may be required from a system development project. Annex E provides details of the technical reviews and audits that could be expected throughout the system development effort. Annex F describes the activities needed to fully define the end products. Annex G provides an explanation of the relationship between the many sources of requirements and the eventual solution.

When the EIA disbanded in 2011, its standards were transferred to other standards organizations.

EIA-632A was superseded by SAE1001 Integrated Project Processes for Engineering a System [4], which replaced EIA-632 when published on 9 October 2018. Although SAE1001 is a successor in one way to EIA-632, the name is changed to reflect that the standard is significantly different. The Society of Automotive Engineers now maintains SAE/EIA-632A-2021, which has been stabilized and is not expected to change for the foreseeable future. SAE1001 is discussed in more detail after discussion of ISO/IEC/IEEE 15288 in Section 6.1.6.

6.1.6 ISO/IEC/IEE 15288 System and Software Engineering—System Life-Cycle Processes

ISO/IEC/IEEE 15288 was ratified as an international standard in mid 2002 and was subsequently revised in 2008, 2015, and 2023. Consideration of ISO/IEC/IEEE 15288 began in the mid 1990s with the intent of developing a standard that provided a common framework within which to address technical (both hardware and software) and social aspects of a system. (In short, it is meant to be able to be applied to any human-made system.) The new standard was to establish common terminology and to integrate across the organization/enterprise all necessary disciplines, technical processes, and project management. Arnold and Lawson [5] provide an excellent summary of the intended role of ISO/IEC/IEEE 15288 from a business management perspective. While earlier standards tended to focus on the design and development aspects of a system life cycle, ISO/IEC/IEEE 15288 defines a set of 30 tailorable processes that covers the system life cycle from a whole-of-business perspective. Again, as with EIA-632, ISO/IEC/IEEE 15288 cannot be considered as a systems engineering standard at all—it certainly does not define systems engineering. As its name implies, the standard is focused on the life cycle of the system, not the project that brings the system into being.

As shown in Figure 6.2 [6], the 30 processes of ISO/IEC/IEEE 15288 rightly have an emphasis on the acquisition phase of the system life cycle, but the whole life cycle is addressed. Figure 6.3 [7] lists the processes in four groups:

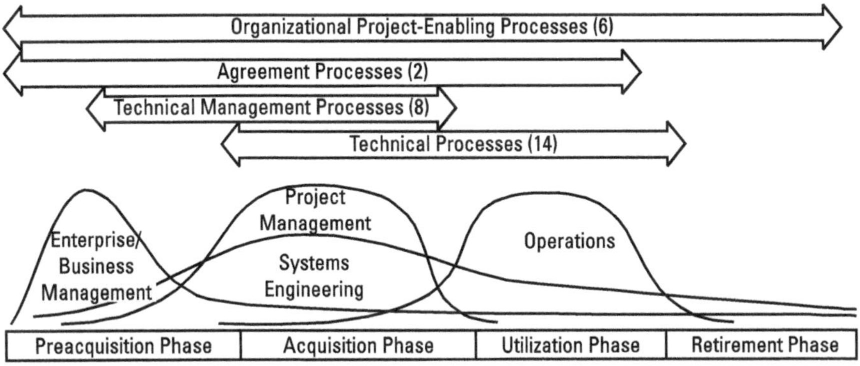

Figure 6.2 The 30 ISO/IEC/IEEE 15288 processes mapped onto the system life cycle.

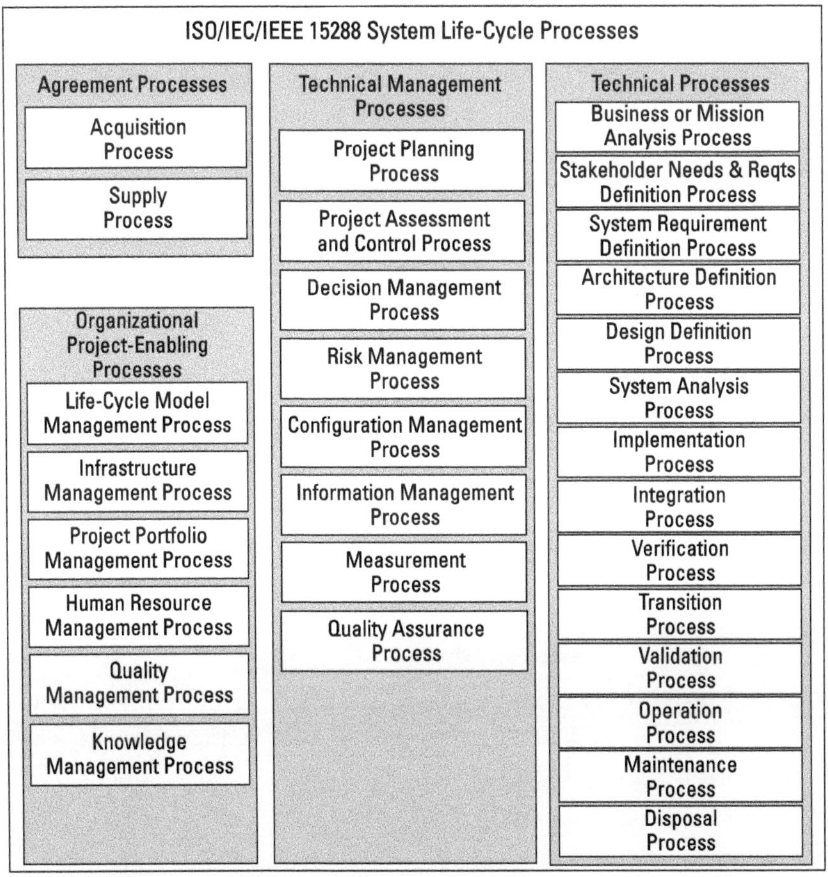

Figure 6.3 ISO/IEC/IEEE 15288 system life-cycle processes.

agreement processes, organizational project-enabling processes, technical management processes, and technical processes.

ISO/IEC/IEEE 15288 establishes a common framework to describe the typical processes associated with the life cycle of any human-made system, which includes anything that "…may be configured with one or more of the following: hardware, software, data, humans, processes (for example, processes for providing service to users), procedures (such as operator instructions), facilities, materials, and naturally occurring entities" [8].

The standard is intended to apply to organizations, or groups of organizations, in their roles as acquirers or suppliers. It is also meant to apply to "… a wide variety of systems in terms of their purpose, domain of application, complexity, size, novelty, adaptability, quantities, locations, life spans, and evolution … to one-of-a-kind systems, mass-produced systems and customized, adaptable systems … complete stand-alone systems and to systems that are embedded and integrated into larger and more complex and complete systems" [9]. In other words, the standard is intended to apply to any system created by humans.

ISO/IEC/IEEE 15288 has a broad application and has been accepted to be the foundational basis of the following standards and extensions:

- ISO/IEC/IEEE 12207: Systems and Software Engineering—Software Lifecycle Processes;
- ISO/IEC/IEEE 15288.1: Application of Systems Engineering for Defense Programs;
- ISO/IEC/IEEE 15288.2: Application of Technical Reviews and Audits for Defense Programs;
- ISO/IEC/IEEE 21839: System of Systems Consideration in the Lifecycle Stages of a System;
- ISO/IEC/IEEE 21840: Systems and Software Engineering—Guidance for the Utilization of ISO/IEC/IEEE 15288 in the Context of System of Systems;
- ISO/IEC/IEEE 21841: Systems and Software Engineering—Taxonomy of System of Systems;
- ISO/IEC/IEEE 24641: Systems and Software Engineering – Methods and Tools for Model-based Systems and Software Engineering;
- ISO/IEC 90005: Guidelines for the Application of ISO 9001 to Systems Lifecycle Processes;
- SAE1001: Integrated Project Processes for Engineering a System.

ISO/IEC/IEEE 15288 is supported by the following guides:

- ISO/IEC/IEEE 24748-1: Systems and Software Engineering—Lifecycle Management – Guidelines for Lifecycle Management;
- ISO/IEC/IEEE 24748-2: Systems and Software Engineering —Lifecycle Management—Guidelines for the Application of ISO/IEC/IEEE 15288;
- ISO/IEC/IEEE 24748-3: Systems and Software Engineering—Lifecycle Management—Guidelines for the Application of ISO/IEC/IEEE 12207;
- ISO/IEC/IEEE 24748-4: Systems and Software Engineering—Lifecycle Management—Application and Management of the Systems Engineering Process (formerly the combination of ISO/IEC 26702: Systems Engineering—Application and Management of the Systems Engineering Process and IEEE-STD 1220: Management of the Systems Engineering Process);
- ISO/IEC/IEEE 24748-5: Systems and Software Engineering—Lifecycle Management—Software Development Planning;
- ISO/IEC/IEEE 24748-6: Systems and Software Engineering—Lifecycle Management—System Integration Engineering;
- ISO/IEC/IEEE 24748-7: Systems and Software Engineering—Lifecycle Management—Application of Systems Engineering on Defense Programs;
- ISO/IEC/IEEE 24748-8: Systems and Software Engineering—Lifecycle Management—Technical Reviews and Audits on Defense Programs.

6.1.7 SAE1001: Integrated Project Processes for Engineering a System

SAE1001: Integrated Project Processes for Engineering a System [11] began as Revision A to EIA-632 and superseded that standard when published on October 9, 2018. Although a successor in one way to EIA-632, the name was changed to reflect that the standard is significantly different.

The current version of SAE1001 is compatible with ISO/IEC/IEEE 15288 (with a mapping provided in Annex B); it uses ISO/IEC/IEEE 15288 terminology and lists ISO/IEC/IEEE 15288 as a normative standard, pointing to definitions and clauses in the higher standard. Although the two standards are ostensibly similar by virtue of their focus on systems engineering processes, SAE1001 is different in that it is focused on a set of "integrated" processes that inform a given project how to implement project processes for engineering a system, providing "…an integrated set of fundamental technical processes to aid a project in the engineering or reengineering of a system." SAE1001 therefore contains details that are not provided by ISO/IEC/IEEE 15288, providing the next level of tasking detail performed by a project, focusing on tasks, inputs, and outputs to show they are interconnected (integrated) to form project processes for systems engineering. That is particularly useful for a small organization that doesn't want the full set of processes as defined in ISO/IEC/IEEE 15288 (particularly the high-level organizational project-enabling processes) but that just wants to focus on the project level.

SAE1001 is therefore targeted at organizations inexperienced with systems engineering that will most likely have immature processes and developing systems engineering personnel so that they are dependent on other organizations for such support. Consequently, the goal of SAE1001 is to assist organizations to improve their systems engineering processes into a cohesive integrated set and to develop their systems engineering staff.

6.1.8 Other Useful Standards

Section 6.1.8.1 lists and describes some additional sources of specific information relating to systems engineering topics. Most of the standards listed are available on the Internet following a simple search using their standard number.

6.1.8.1 Other Systems Engineering Standards

ECSS-E-10A [12] is part of a set of standards issued by the European Cooperation for Space Standardization and tailored for use in space applications. It provides useful coverage of the requirements engineering process, configuration management, and systems engineering products and interfaces with other disciplines. The standard is written as a set of statements of what must be done in systems engineering rather than describing how to do it.

In 2004 ECSS-E-10A was superseded by ECSS-E-10 Part 1B, reflecting updates and refinements in system engineering practices within the space sector. On 6 March 2009, ECSS-E-10 Part 1B was replaced by ECSS-E-ST-10C, which represented a significant shift by presenting only requirements and deferring informative material to a new handbook, ECSS-E-HB-10. This change aimed to streamline the standard and provide clearer guidance for its application.

There are many other standards related to systems engineering, including the following:

- ISO/IEC/IEEE 15026-1: Systems and Software Engineering — System and Software Assurance — Part 1: Concepts and Vocabulary;
- ISO/IEC/IEEE 15026-2: Systems and Software Engineering — System and Software Assurance — Part 2: Assurance Case;
- ISO/IEC/IEEE 15026-3: Systems and Software Engineering — System and Software Assurance — Part 3: Integrity Levels;

- ISO/IEC/IEEE 15026-4: Systems and Software Engineering — System and Software Assurance — Part 4: Assurance in the Lifecycle;
- ISO/IEC/IEEE 15289: Systems and Software Engineering — Content of Lifecycle Information Products (Documentation);
- ISO/IEC/IEEE 24765: Systems and Software Engineering — Systems and Software Engineering Vocabulary;
- ISO/IEC/IEEE 16326: Systems and Software Engineering — Life Cycle Processes — Project Management;
- ISO/IEC/IEEE 16085: Systems and Software Engineering — Life Cycle Processes — Risk Management;
- ISO/IEC/IEEE 42010: Software, Systems and Enterprise — Architecture Description;
- ISO/IEC/IEEE 42020: Software, Systems and Enterprise — Architecture Process;
- ISO/IEC/IEEE 42030: Software, Systems and Enterprise — Architecture Evaluation Framework;
- ISO/IEC/IEEE 29148: Systems and Software Engineering — Requirements Engineering;
- ISO/IEC/IEEE 15939: Systems and Software Engineering — Measurement Process;
- ISO/IEC/IEEE 26560: Systems and Software Engineering — Reference Model for Product Line Engineering and Management;
- ISO/IEC/IEEE 26580: Systems and Software Engineering — Models and Tools for the Feature-based Approach to Software and Systems Product Line Engineering;
- ISO 10303-233: Systems Engineering Data Interchange Standard;
- ISO/IEC 29110: System and Software Engineering Standards and Guides for Very Small Entities;
- ISO/IEC 31000: Risk Management;
- ISO/IEC 31010: Risk Management — Risk Assessment Techniques;
- ISO/IEC 33060: Process Assessment — Process Assessment Model for System Life Cycle Processes;
- NIST SP 800-160 Vol. 1: Systems Security Engineering: Considerations for a Multidisciplinary Approach in the Engineering of Trustworthy Secure Systems;

- NIST SP 800-160 Vol. 2: Developing Cyber-Resilient Systems: A Systems Security Engineering Approach;
- Capability maturity models (CMMs) such as those administered by the CMMI Institute;
- OMG SysMLTM: OMG Systems Modeling Language.

6.1.8.2 Technical Reviews and Audits

MIL-STD-1521B [13] is an old standard that was canceled by the US Department of Defense on April 10, 1995. Some of its content was summarized and divided between MIL-STD-499B and MIL-STD-973. Despite this, it remains an excellent reference source for technical reviews and audits—a quick scan will recognize all the reviews and audits we have mentioned here: SRR, SDR, PDR, CDR, TRR, FCA, and PCA. It provides thorough coverage of a comprehensive range of reviews and audits and links the reviews and audits to the systems engineering life cycle presented in the systems engineering standard of the day.

MIL-STD-1521B provides useful descriptions of the types of issues to be covered at each of the reviews and the criteria that should be met prior to entering and exiting each review. In addition, it provides checklists and templates that may be of assistance during the technical review and audit program.

In defense programs, the requirements for conducting reviews and audits are contained in IEEE 15288.2-2014 IEEE Standard for Technical Reviews and Audits.

6.1.8.3 Configuration Management

EIA-649C [14], MIL-STD-973 [15], and MIL-HDBK-61B [16] are three standards that describe CM practices and processes described in Chapters 2 to 5. MIL-STD-973 is generally regarded as the classic configuration management standard and remains an excellent source of reference, although it is becoming a little dated in some regards and was canceled with the release of EIA-649C. More recently, ISO 10007 [17] provides guidelines for configuration management.

MIL-HDBK-61B [18] is a military handbook on configuration management. Like other military handbooks, MIL-HDBK-61B is not designed to be contractually invoked but is, instead, written as a guidance document. Regardless of what configuration management standard is being used, MIL-HDBK-61B is valuable source of information about modern configuration management practice.

6.1.8.4 Specification Standards

MIL-STD-490A [19] was canceled and replaced by MIL-STD-961D [20]. However, as is often the case, MIL-STD-490A remains a source of information on specifications. The standards introduce a hierarchical set of specifications from the A-level system specification to the E-level materials specification. The relationship of these standards to functional (A-level specification), allocated (B-level specifications), and product baselines (C-, D-, and E-level specifications) is also explained. The standard also describes the many different types of development and product specifications including prime item and critical item development and product specifications. The document provides content information for each of the specifications in an appendix.

MIL-STD-961D was written to describe the content and requirements of generic specifications rather than content of specific specifications as outlined in its predecessor. With that in mind, reference to both types of specifications can assist in understanding of the nature and content of technical specifications required on given system developments.

6.1.8.5 WBSs

MIL-STD-881E, Work Breakdown Structures for Defense Materiel Items [21], is an excellent source of information on work breakdown structures effectively describes the relationship between the program WBSs (PWBSs) and the contract WBSs (CWBSs). It also contains detailed information on the purpose, structure, and content of a range of example programs including an aircraft system that proved useful in this text.

6.1.9 Useful Guidance from INCOSE

INCOSE is a nonprofit membership organization founded to develop and disseminate the transdisciplinary principles and practices that enable the realization of successful systems [22]. In addition to connecting systems engineering professionals with educational, networking, and career-advancement opportunities, INCOSE is also focused on producing state-of-the-artwork products that support and enhance systems engineering, providing standards, handbooks, and guidance for professionals across industries. There are many such products available in the INCOSE store (free to INCOSE members and at a small cost for nonmembers) that conform to the overarching ISO/IEC/IEEE 15288; this section briefly describes the major handbooks and guides as follows.

- *Systems Engineering Handbook* [23]. The INCOSE *Systems Engineering Handbook* is an essential reference for systems engineering practitioners and engineers in other disciplines looking to perform or understand the discipline of systems engineering. The *Handbook* elaborates on the key systems life-cycle processes described in ISO/IEC/IEEE 15288 with chapters covering key systems engineering concepts, system life-cycle processes and methods, tailoring and application considerations, systems engineering in practice, and more. The *Handbook* is the basis for certification of the systems engineering discipline.

- *Guide to the Systems Engineering Body of Knowledge* (SEBoK) [24]. The SEBoK is an online resource that provides a guide to the key knowledge sources and references of systems engineering, organized and explained to assist a wide variety of individuals. It is a living product, accepting community input continuously, with regular updates.
- *Needs and Requirements Manual* (NRM) [25]. This INCOSE manual is intended for those whose role is to perform needs, requirements, verification, and validation activities throughout the system life cycle and is addressed to practitioners of all levels of experience. Those new to project management and system development should find the detailed guidance useful. Those more experienced should be able to find new insights across all phases of the life cycle, particularly in terms of a data-centric perspective to which they may not have been exposed. The NRM is supported by several INCOSE guides:
    - *Guide to Needs and Requirements* [26];
    - *Guide to Writing Requirements* [27];
    - *Guide to Verification and Validation* [28];
    - *Guide to Security Needs and Requirements.*
- Other useful INCOSE artifacts:
    - *Integrated Data as a Foundation of Systems Engineering;*
    - *Systems Engineering Tools Database (SETDB);*
    - *A Complexity Primer for Systems Engineers;*
    - *Digital Engineering Measurement Framework;*
    - *Guide for the Application of Systems Engineering in Large Infrastructure Projects;*
    - *SE Competency Framework;*
    - *Project Managers Guide to SE Measurement for Project Success;*
    - *Systems Engineering Agility Primer;*
    - *Systems Engineering Measurement Primer;*
    - *Systems of Systems Primer;*
    - *Systems Engineering Principals;*
    - *Technical Measurement Guide.*

6.2 Systems Engineering Process Tools

This section briefly examines tools to assist in the conduct of systems engineering processes. Chapters 3 and 4 cover a number of analysis tools; this chapter focuses on schematics, modeling, and simulation as tools that can assist the synthesis aspect of the systems engineering processes. Trade-off analysis is also detailed as a potential tool for use during the evaluation stage of the process.

6.2.1 Analysis Tools

Chapters 2 to 4 describe some analysis tools, including the context diagram, FFBD, RBS and N2 Diagrams. Other tools include structured analysis, data flow diagrams (DFDs), control flow diagrams (CFDs), IDEF diagrams, behavior diagrams, action diagrams, state/mode diagrams, process flow diagrams, function hierarchy diagrams, state transition diagrams, entity relationship diagram (ERDs), structured analysis and design, object oriented analysis (OOA), unified modeling language (UML), structured systems analysis and design (SSADM), and quality function deployment (QFD). Each of these techniques focuses on gathering and exploring requirements in a formal systematic way, which is commonly referred to generically as business analysis [29] or requirements engineering [30]. Chapters 2 and 3 focus on the philosophy of requirements engineering, since it is an essential element of system development; we leave the reader to consult the many excellent references for the more detailed tools [31].

6.2.2 Synthesis Tools

There are several tools available for use during the synthesis stage of the systems engineering process. This section aims to introduce the more common tools and briefly explain their characteristics.

6.2.2.1 Block Diagrams

Block diagrams (or simply schematics) are perhaps the most basic and therefore popular of the synthesis tools. Schematics allow designers to visually describe a solution to a given functional need as defined during the analysis stage of the process. Schematics show all the elements of the solution along with their interrelationships and interfaces. These interfaces include internal interfaces (between schematic elements) and external interfaces (with other systems or subsystems). Schematics can be drawn at different levels of detail during systems engineering processes. For example, high-level schematics drafted during preliminary design will be far less detailed than the schematics used to assist with the detailed design and development of subsystem elements.

An example of an avionics system schematic block diagram is shown in Figure 6.4, illustrating the major subsystem elements, their interrelationships and internal interfaces, and external interfaces. Notice how readily the design and operation of this avionics subsystem can be understood.

A major advantage of schematic block diagrams is their ability to clearly communicate the proposed solution to technical and nontechnical persons alike—a picture is worth a thousand words.

6.2.2.2 Physical Modeling

Physical modeling can take one of several forms. Perhaps the most well-known of physical models are engineering models such as the wind tunnel models used during vehicle design to determine aerodynamic stability and characteristics. These models can include models of aircraft, land vehicles, and sea-going vehicles such as ships and submarines. They are normally scale models and are built to exacting standards to ensure that the results obtained during the trials will be representative of the full-scale variant. Full-scale mockups of vehicles are also sometimes built to allow human factors engineers to experiment with design decisions such as cable routing, access points, and operator equipment. Electrical and electronic design can also benefit from physical modeling. Examples include breadboards of electrical and electronic circuits and brassboards used during the design of RF elements. These mockups are also examples of physical models.

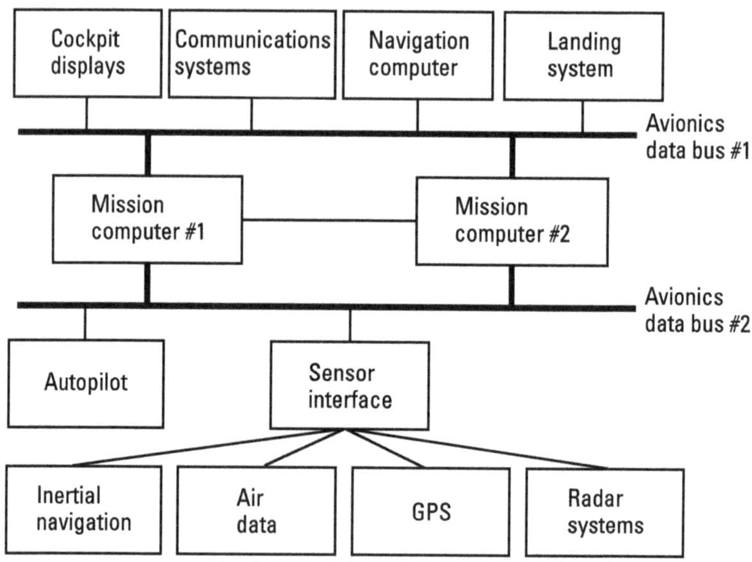

Figure 6.4 Example schematic block diagram.

It is fair to say that the use of physical models in today's design and development environment is being reduced as the power and capability of computer-based modeling and simulation tools increases. However, sophisticated computer-aided tools become, physical modeling will still play a role in the synthesis of certain design solutions.

6.2.2.3 Mathematical Modeling and Simulation

Mathematical modeling and simulation are widely used and becoming increasingly popular as computers and software tools become more capable. As the name suggests, mathematical models rely on a mathematical representation of a physical system or design. The models can then be used to determine both static and dynamic characteristics of the system. An example of a static model would be one that represents a bridge's structural characteristics. Loads and stresses can be changed in magnitude and position to ensure that the structure will withstand the loads associated with its intended purpose. An example of a dynamic system is an aircraft's autopilot system. This is a classic example of a control system. Characteristics such as stability and response times can be determined using this sort of model.

Finite element analysis (FEA) [32], a very well-known mathematical modeling technique, serves as a good example of how mathematical modeling can assist and enhance synthesis throughout the systems engineering process. FEA is a mathematical technique used to study the static and dynamic behavior of physical structures such as aircraft wings when they are placed under varying conditions. FEA is used to understand this behavior without the need for expensive and time-consuming prototypes and tests. The structure under consideration is assumed to consist of a finite arrangement of individual elements. Each element is modeled mathematically using relatively straightforward physical and engineering laws. The individual models are then combined in such a way as to represent and model the entire structure. The model of the entire structure is analyzed to assist the iterative design (synthesis) of the structure.

FEA assists by ensuring that the design of the structure adequately meets the requirements assigned to it during the systems engineering process. It also ensures that the structure is not overengineered unnecessarily, saving valuable time and money during design.

Models and simulations are excellent tools for performing what-if analysis and determining the design limits of the design. Variables can be changed and conditions altered easily to allow designers to investigate the likely behavior of the design in varying conditions. Simulators are widely used for a variety of different synthesis tasks. Environmental design and HMI design are two examples of synthesis tasks that can benefit from simulation. The fidelity of simulators must be defined and confirmed prior to their use as synthesis tools.

To build and use a mathematical model, designers need to define the problem and formulate the model to replicate the problem. This model can be used with pen and paper if desired, but the more common approach is to convert the model into a computer program for ease of use and speed. Once this is done, perhaps the most crucial step must be performed. The performance of the model must be verified to ensure that the model is an accurate representation of the problem. It is pointless collecting data from a model if the model is incorrect (not representative). Once the model has been built and verified, designers can experiment with it and collect important design data.

6.2.3 Evaluation Tool—Trade-Off Analysis

Throughout the systems engineering management and processes chapters of this text, references are made to the evaluation of alternative solutions and the selection of the preferred approach. The process of analyzing alternatives and selecting the preferred solution is often generically called trade-off analysis. The analysis process needs to consider the different alternatives and balance the many (potentially conflicting) requirements prior to selecting the preferred solution. It must be emphasized that trade-off analysis is a decision-making aid only; it is not the decision maker.

6.2.3.1 Analysis Utility

Trade-off analysis is used to assist decision making throughout all phases of the system life cycle. Table 6.2 summarizes typical activities conducted during the acquisition phase that may benefit from the use of trade-off analysis [33].

6.2.3.2 Analysis Process

The process employed in conducting a trade-off analysis will vary with the individual needs of the decision-making process. This section provides a guide

Table 6.2
Trade-Off Analysis Examples

Acquisition Phase	Potential Applications
Conceptual Design	Investigation of candidate new and emerging technology Selection of the preferred system design concept Identification of potential system configurations
Preliminary Design	Selection of the technology to be employed Selection of the preferred system configuration
Detailed Design	Consideration/selection of subsystem and component design Consideration/selection of testing methodology
Construction/Production	Consideration of design changes Optimization use of design space

to the general process that may be employed. Steps in the analysis are described as follows:

- *Definition of requirements.* The requirements and needs of the problem to be investigated need to be recorded in precise terms. Systems engineering artifacts such as specifications and design aids such as function flow block diagrams will provide a good source of this information. Sources will vary as the development moves through the system life cycle.
- *Identification of alternative solutions:* The trade-off analysis must identify the widest range of alternative solutions possible to maximize the chance of selecting the optimal solution to the problem. There are many sources of alternative solutions including systems engineering synthesis activities, OTS equipment, and other existing equipment. Once the range of solutions has been identified, they should pass through a screening process to eliminate infeasible or unsuitable solutions from further consideration. This will avoid investing valuable time and effort producing solutions that ultimately cannot be implemented for one reason or another. The remaining (feasible) solutions should be described in sufficient detail to allow further detailed consideration at a later stage in the process.
- *Nomination of selection criteria:* Criteria must be formulated to provide a measure of the effectiveness of the alternatives to meet the requirements of the problem. If possible, these measures should be quantitative in nature to avoid, as much as possible, subjective assessment of the alternatives. The different criteria should be independent from one another to avoid assessing the same characteristics a number of times, and the criteria should be directly related to the defined problem.
- *Determination of criteria weighting:* Once the criteria have been identified, they should be weighted according to their relative importance. Some criteria may be considered equally important as one another and be assigned the same weighting; however, in most cases, there will be a clear difference in the level of importance among the criteria. This should be reflected in weighting. If considered necessary, the weightings may be withheld from the individuals performing the evaluation to help ensure the objectivity of the assessment. Numerical assessment using weightings will assist in the comparative assessment of the alternatives under consideration. By combining an assessment of how well an alternative meets a certain criterion with a measure of the importance of that criterion, the evaluation will provide an excellent indication of the alternative's performance.

- *Scoring function:* It may be worthwhile to assign scoring functions to each criterion to ensure that the performance of each of the alternatives against the criteria is scored consistently. The scoring functions will be used by the scorers during the assessment of each alternative. A range of scores should be selected (for example from 0 to 10) and this range applied to all criteria. By applying different scoring ranges to different criteria, inadvertent weighting will be introduced. Weighting has already been assigned during the previous step.

Example 6.1: Aircraft System Weighting

Figure 6.5 illustrates criteria weighting for our aircraft example.

Example 6.2: Aircraft System Scoring Function

Using the aircraft example, the scoring function in Figure 6.6 may be applicable to the cruising speed criterion.

In both examples, candidate aircraft that have a cruising speed of greater or equal to 500 knots will score 10 points. Speeds of less than 300 knots will be given a score of 0. In the linear example on the left, a score is awarded in accordance with a linear interpolation for speeds between 300 and 500 knots which scores 10—speeds greater than 500 knots score no greater than 10. In the step-function example on the right, a higher score of 5 is available for speeds between 300 and 400 knots, a score of 10 for speeds greater than 500 knots, but no score greater than 10 for speeds greater than 500 knots.

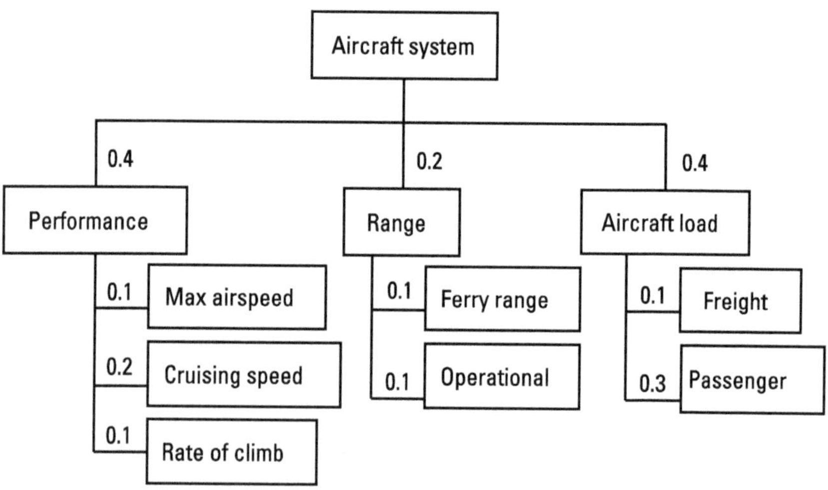

Figure 6.5 Example aircraft system weighting.

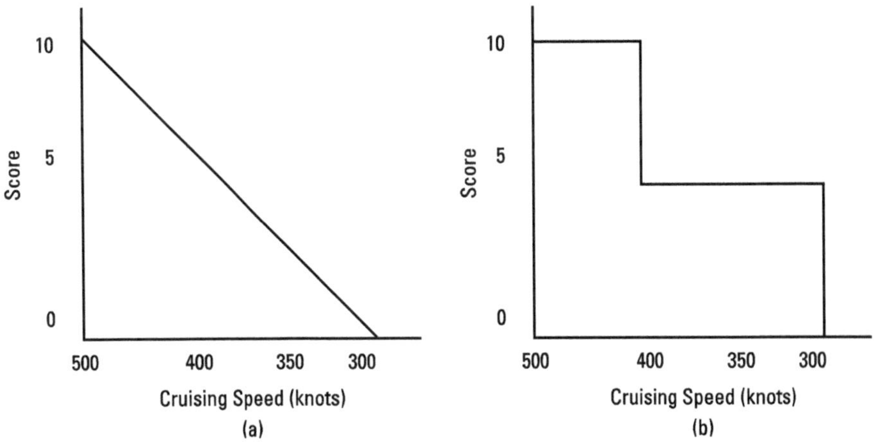

Figure 6.6 Illustration of a scoring function for the cruising speed criterion for our aircraft system: (a) linear and (b) step function.

- *Evaluation of alternatives:* Once all the weights and scoring functions have been produced, the alternatives can be assessed. Normally the scores and weightings will be recorded in a tabulated form and the total score for each alternative will be calculated using the following (or similar) expression:

$$\text{Weighted score} = \sum_{i=1}^{N} w_i s_i$$

Where N is the number of criteria, w_i is the weighting of ith criteria, and s_i is the score of ith critiera.

Provided the criteria, weightings and scoring functions have been correctly compiled, the weighted scores will then provide an indication of the relative order of merit of each of the alternatives.

- *Sensitivity study:* A sensitivity study, or sensitivity analysis, should be performed to confirm the validity of the results of the overall trade-off analysis. The sensitivity check will concentrate on the range of possible scores that could have been achieved by each alternative against each criterion (given that the score assessment may have been subjective to an extent). The range of scores will then translate into a range of weighted scores for each alternative. Sensitivity studies attempt to avoid the situation where the incorrect alternative is selected due to poorly informed assessment by investigating the situation where a small change in a single score can have significant effect on the overall weighted score. If the score ranges still clearly identify the preferred alternative, the sensitivity study will have reinforced the validity of that alternative. If, on the other hand, the score ranges overlap, the decision makers need to consider the following courses of action, listed as follows:

- Obtain additional information for each alternative to further differentiate between them.
- Review the criteria, weightings, and scoring functions to ensure that they are correct.
- If the results are still inconclusive, proceed with caution until evidence becomes available to allow a clear differentiation among alternatives.

6.3 MBSE

Section 6.2.2 discusses the use of modeling and simulation and provides a definition of a model. When modeling becomes the central part of a systems engineering process, the process becomes known as MBSE [34]. INCOSE defines MBSE as, "...the formalized application of modelling to support system requirements, design, analysis, verification and validation activities beginning in the conceptual design phase and continuing throughout development and later life cycle phases" [35].

MBSE is often contrasted with what some call "document-based" systems engineering. In document-based systems engineering, critical systems engineering information is captured in artifacts like the ones described in this text. For example, system-level requirements are captured in the SyRS, interface information captured in ICDs, design information captured in the SDD, verification procedures in planning documents, and so on. In an MBSE process, this information is contained in a model (or models). Stakeholders view the information they are interested in by using an array of views (or diagrams) extracted from that model. For example, the analysis and information contained in a traditional schematic block diagram (Section 6.2.2.1) can be developed and viewed in MBSE diagrams like the SysML Block Definition Diagram (BDD) and the Internal Block Diagram (IBD) available within standards-based MBSE tools. The model therefore becomes the single source of truth, because the same model is being viewed regardless of whether the stakeholder is interested in system requirements, interface requirements, system design, or verification. If a stakeholder detects an error in their particular view, they can correct the model, and that correction becomes available immediately to the other stakeholders who are using the model for their own purposes.

When used successfully, MBSE offers the potential to improve stakeholder decision-making and understanding; enhance communications; support trade-off analyses; and capture system behavior, structure, and requirements across all levels of the system hierarchy. MBSE also promises to allow systems engineers to measure progressively the completeness and correctness of their work by detecting errors and omissions in the systems engineering effort. MBSE may also enhance the ability of systems engineers to deal with change management efficiently and effectively as part of a wider configuration management process.

Adopting an MBSE approach requires stakeholders to agree on three fundamental things: the underlying systems engineering process to be used, the MBSE toolset to be used, and the formal MBSE language to be used in the modelling effort. Standard systems engineering processes such as the one described in this book can be adopted to MBSE, but commercially available MBSE process models also exist. These commercially available processes are generally associated with a particular MBSE toolset that usually must be purchased before it can be used. For example, IBM has published its Harmony methodology, which is designed to be used in conjunction with their MBSE tool called Rhapsody. The formal language that is used by a majority of the MBSE toolsets is called System Modeling Language (SysML) [36] and is generally considered to be the default standard MBSE language. The formal language is critical because it determines views and diagrams, the notation and syntax to be applied, and rules associated with both static and dynamic models of the system. Using a standardized language also allows for sharing of models across different organizations, different organizational levels, and potentially across MBSE toolsets from different vendors.

As in more traditional systems engineering processes, the model at the heart of the MBSE process can be developed progressively: from operational to component level, from functional to physical views, and across various levels of a system hierarchy via a standard systems engineering life cycle. Alternatively, organizations may start the systems engineering process for a new or improved system by modeling the current system that needs to be replaced or improved.

As organizations around the world continue to develop their systems engineering processes and adopt concepts such as MBSE, they are experiencing and overcoming common challenges. For example, to interact with and understand the model(s), expertise is required in both the modeling language (such as SysML) and in the selected MBSE tool. While this may not be a challenge for an MBSE practitioner, it creates a communication and comprehension challenge for a host of other technical and nontechnical stakeholders for whom the modeling may have been performed. Effort is required to become sufficiently proficient in both the language and the toolset to understand the different views and usages of the model. This effort is not just time-related, because commercially available tools require additional financial investment in terms of both licensing and training costs. An integrated MBSE approach to system development can also be difficult in a classic acquirer/developer procurement model, especially if there are significant differences in how the acquirer and developer are performing and using MBSE. The same challenge is replicated within a developer's organization in the handover of systems engineering work where there are differences between how the systems engineers are working and how the more traditional detailed design and development disciplines are working.

Excellent coverage of MBSE and its application can be found in several texts [37].

6.4 Summary

This chapter reviews several relevant standards and discusses their utility to the systems engineering management effort. We also summarize the major systems engineering guidance available from INCOSE, which is freely available to their members.

While there has been considerable movement toward a harmonious suite of standards in ISO/IEC/IEEE 15288 and its attendant handbooks, there is some way to go and there are still several organizations around the world making use of older standards such as EIA-632 and IEEE-STD-1220. There are also many organizations whose internal processes have been developed based on an older standard—there is considerable inertia to rapid change. Having said that, while there are clear differences in the various standards, there are many more similarities, and they all provide a framework for the establishment of useful systems engineering processes and procedures in an organization. In any case, any selected standard will almost always have a scope greater than the project and system under consideration—so tailoring is essential, not only because of the scope of the standard but also because of the need to temper the selected standard's requirements with the culture and structure of the adopting organization.

We also briefly discuss several systems engineering process tools that support analysis, synthesis, and evaluation, and provided an overview of MBSE.

Review Questions

1. Explain the systems engineering processes used by MIL-STD-499B, IEEE-STD-1220, and EIA/IS-632.
2. Explain how systems engineering standards could be used to assist a systems engineer during the planning of stages of a developmental project.
3. Explain how systems engineering standards could assist an acquirer in defining systems engineering requirements in a contract.

4. Briefly explain the differences between the earlier systems engineering standards (such as MIL-STD-499B, IEEE-STD-1220, and EIA/IS-632), and later standards such as EIA-632, ISO/IEC/IEEE 15288, and SAE 1001.
5. Briefly describe the seven steps of a trade-off analysis.
6. Briefly describe MBSE and its utility.

Endnotes

[1] EIA-632, *Processes for Engineering a System,* Washington, DC: Electronic Industries Association (EIA), 2003.

[2] Martin, J., "Processes for Engineering a System: An Overview of the EIA-632 Standard and Its Heritage," *Systems Engineering,* Vol. 3, No. 1, 2000, pp. 1-26.

[3] EIA-632, *Processes for Engineering a System,* Washington, DC: Electronic Industries Association (EIA), 2003.

[4] SAE1001, *Integrated Project Processes for Engineering a System,* SAE International, 2018.

[5] Arnold, S., and H. W. Lawson, "Viewing Systems from a Business Management Perspective: The ISO/IEC 15288 Standard," *Systems Engineering,* Vol. 7, No. 3, 2004, pp. 224–242.

[6] ISO/IEC/IEEE 15288, *Systems Engineering—System Lifecycle Processes,* International Organization for Standardization, 2023.

[7] ISO/IEC/IEEE 15288, *Systems Engineering—System Lifecycle Processes,* International Organization for Standardization, 2023.

[8] ISO/IEC/IEEE 15288, *Systems Engineering—System Lifecycle Processes,* International Organization for Standardization, 2023.

[9] ISO/IEC/IEEE 15288, *Systems Engineering—System Lifecycle Processes,* International Organization for Standardization, 2023.

[10] SAE1001, *Integrated Project Processes for Engineering a System,* SAE International, 2018.

[11] SAE1001, *Integrated Project Processes for Engineering a System,* SAE International, 2018.

[12] ECSS-E-10A, *Space Engineering—System Engineering,* Noordwijk, The Netherlands: European Cooperation for Space Standardization, 1996.

[13] MIL-STD-1521B, *Military Standard—Technical Reviews and Audits for Systems, Equipments, and Computer Software,* Washington, DC: US Department of Defense, 1985.

[14] EIA-649C, *Configuration Management Standard,* Warrendale, PA: SAE International, 2019.

[15] MIL-STD-973, *Military Standard—Configuration Management,* Washington, DC: US Department of Defense, 1992.,

[16] MIL-HDBK-61B, *Military Handbook—Configuration Management Guidance,* Washington, DC: US Department of Defense, 2020.

[17] ISO 10007, *Quality Management Systems—Guidelines for Configuration Management,* 2017.

[18] MIL-HDBK-61B, *Military Handbook—Configuration Management Guidance,* Washington, DC: US Department of Defense, 2020.

[19] MIL-STD-490A, Military Standard—Specification Practices, Washington, DC: US Department of Defense, 1985.

[20] MIL-STD-961E, *Department of Defense Standard Practice for Defense Specifications,* Washington, DC: US Department of Defense, 2008.

[21] MIL-STD-881E, *Department of Defense Standard Practice —Work Breakdown Structures for Defense Material Items,* Washington, DC: US Department of Defense, 2020.

[22] www.incose.org.

[23] International Council on Systems Engineering (INCOSE), INCOSE-TP-2003-002-05, *Systems Engineering Handbook,* Fifth Edition, New York: John Wiley & Sons, 2023.

[24] SEBoK (2024), Guide to the Systems Engineering Body of Knowledge (SEBoK), https://sebokwiki.org/wiki/Guide_to_the_Systems_Engineering_Body_of_Knowledge_(SEBoK).

[25] Wheatcraft, L. S., M. J. Ryan, and T. E. Katz, *Needs and Requirements Manual: Needs, Requirements, Verification, Validation Across the Lifecycle,* Hoboken, NJ: John Wiley & Sons, 2024.

[26] International Council on Systems Engineering (INCOSE), INCOSE-TP-2021-003-01, *Guide to Needs and Requirements,* 2022.

[27] International Council on Systems Engineering (INCOSE), INCOSE-TP-2010-004-04, *Guide to Writing Requirements,* 2024.

[28] International Council on Systems Engineering (INCOSE), INCOSE-TP-2021-006-01, *Guide to Verification and Validation,* 2022.

[29] International Institute for Business Analysis, *A Guide to the Business Analysis Body of Knowledge® (BABOK® Guide),* Version 3, 2015.

[30] In addition to information contained in the general systems-engineering references listed in Chapter 1's endnotes, specific detail on requirements engineering can be found in the following:

Alexander I. F., and L. Beus-Dukic, *Discovering Requirements: How to Specify Products and Services,* Chichester, England: John Wiley & Sons, 2007.

Andriole, S., *Managing Systems Requirements: Methods, Tools and Cases,* New York: McGraw-Hill, 1996.

Aurum, A., and C. Wohlin (eds.), *Engineering and Managing Software Requirements,* Berlin: Springer, 2005.

Berenbach, B., et al., *Software and Systems Requirements Engineering in Practice,* New York: McGraw-Hill, 2009.

Bray, I. K., *An Introduction to Requirements Engineering,* Boston, MA: Addison Wesley, 2002.

Bron, J-Y., *System Requirements Engineering: A SysML Supported Requirements Engineering Method,* Wiley-ISTE, 2020.

Fernandes, J., and R. Machado, *Requirements in Engineering Projects,* Cham, Switzerland: Springer, 2016.

Gilb, T., *Competitive Engineering: A Handbook for Systems Engineering, Requirements Engineering, and Software Engineering Using Planguage,* Amsterdam: Butterworth-Heinemann, 2005.

Goldsmith, R. F., *Discovering Real Business Requirements for Software Project Success,* Norwood, MA: Artech House, 2004.

Grady, J., *Systems Requirements Analysis,* New York: McGraw-Hill, 2006.

Hass, K. B., D. J. Wessels, and K. Brennan, *Getting it Right: Business Requirement Analysis Tools and Techniques,* Vienna, VA: Management Concepts, 2008.

Hay, D. C., *Requirements Analysis: From Business Views to Architecture,* Upper Saddle River, NJ: Prentice Hall, 2003.

Hood, C., et al., *Requirements Management: The Interface Between Requirements Development and All Other Systems Engineering Processes,* Oberhaching, Germany: Springer, 2008.

Hooks, I. F., and K. A. Farry, *Customer-Centered Products: Creating Successful Products Through Smart Requirements Management,* New York: AMACOM, 2001.

Hossenlopp, R., and K. B. Hass, Unearthing Business Requirements: Elicitation Tools and Techniques, Vienna, VA: Management Concepts, 2008.

Hull, E., K. Jackson, and J. Dick, *Requirements Engineering,* London: Springer, 2011.

International Institute for Business Analysis, *A Guide to the Business Analysis Body of Knowledge® (BABOK® Guide),* Version 3, 2015.

ISO/IEC/IEEE 29148, *Systems and Software Engineering—Lifecycle Processes—Requirements Engineering,* 2018.

Jackson, M., *Software Requirements and Specifications: A Lexicon of Software Practice, Principles and Prejudices,* Reading, MA: Addison Wesley, 1995.

Jonasson, H., *Determining Project Requirements,* Second Edition, *Mastering the BABOK® and the CBAP® Exam,* Boca Raton, FL: Auerbach Publications, 2012.

Kossmann, M., *Requirements Management: How to Ensure You Achieve What You Need From Your Projects,* Gower, 2013.

Kotonya, G., and I. Sommerville, *Requirements Engineering: Processes and Techniques,* West Sussex, England: John Wiley & Sons, 2000.

Leffingwell, D., *Agile Software Requirements: Lean Requirements Practices for Teams, Programs, and the Enterprise,* Reading, MA: Addison-Wesley, 2011.

Mate J. L., and A. Silva (eds.), *Requirements Engineering for Sociotechnical Systems,* Hershey, PA: Information Science Publishing, 2005.

Pohl, K., *Requirements Engineering: Fundamentals, Principles, and Techniques,* Heidelberg, Germany: Springer-Verlag, 2010.

Pohl, K., *Requirements Engineering Fundamentals,* Santa Barbara CA: Rocky Nook, 2011.

Pohl, K., and C. Rupp, *Requirements Engineering: Fundamentals, Principles, and Techniques,* Heidelberg, Germany: Springer-Verlag, 2011.

Robertson S., and J. Robertson, *Mastering the Requirements Process,* Harlow, England: Addison-Wesley, 2013.

Ryan, M. J., *Requirements Practice in Conceptual Design: A Practical Guide,* Norwood, MA: Artech House, 2026.

Sommerville, I., and P. Sawyer, *Requirements Engineering,* Chichester, England: John Wiley & Sons, 1997.

Thayer, R. H., and M. Dorfman (eds.), *Software Requirements Engineering,* Los Alamitos, CA: IEEE Computer Society Press, 1997.

van Lamsweerde, A., *Requirements Engineering: From System Goals to UML Models to Software Specifications,* Chichester, England: John Wiley & Sons, 2009.

Wheatcraft, L. S., M. J. Ryan, and T. E. Katz, *Needs and Requirements Manual: Needs, Requirements,* Verification, Validation Across the Lifecycle, Hoboken, NJ: John Wiley & Sons, 2024.

Wiegers, K. E., and J. Beatty, *Software Requirements,* Redmond, WA: Microsoft Press, 2013.

Wieringa, R., *Requirements Engineering: Frameworks for Understanding,* New York: John Wiley & Sons, 1996.

Withall, S., *Software Requirements Patterns (Best Practices),* Redmond, MA: Microsoft Press, 2007.

Young, R., *The Requirements Engineering Handbook,* Norwood, MA, Artech House, 2004.

[31] Specific detail on analysis tools can be found in the following:

Blanchard, B., and W. Fabrycky, *Systems Engineering and Analysis,* Upper Saddle River, NJ: Prentice Hall, 2100.

Coad, P., and E. Yourdon, *Object-Oriented Analysis,* Upper Saddle River, NJ: Prentice Hall, 1990

Defense Systems Management College (DSMC), *Systems Engineering Management Guide,* Washington, DC: US Government Printing Office, 1990.

Eisner, H., *Essentials of Project and Systems Engineering Management,* New York: John Wiley & Sons, 2008.

Gause, D., and G. Weinberg, *Exploring Requirements: Quality Before Design,* New York: Dorset House Publishing Company, Incorporated, 1989.

Goodland, M., and C. Slater, *SSADM: A Practical Approach,* London: McGraw-Hill Publishing Company, 1995.

Grady, J., *Systems Requirements Analysis,* New York: McGraw-Hill Book Co., 1994.

Heap, G., J. Stanway, and A. Windsor, *A Structured Approach to Systems Development,* London: McGraw-Hill Book Company, 1992.

Hitchins, D., *Putting Systems to Work,* Chichester, England: John Wiley & Sons, 1992.

Kotonya, G., and I. Sommerville, *Requirements Engineering: Processes and Techniques,* West Sussex, England: John Wiley & Sons, 2000.

Lacy, J., *Systems Engineering Management: Achieving Total Quality,* New York: McGraw-Hill Book Co., 1992.

Leffingwell, D., and D. Widrig, *Managing Software Requirements: A Unified Approach,* Reading, MA: Addison-Wesley, 2000.

Macaulay, L., *Requirements Engineering,* London, England: Springer, 1996.

Maier, M., and E. Rechtin, *The Art of Systems Architecting,* Boca Raton, FL: CRC Press, 2000.

Martin, J., *Principles of Object-Oriented Analysis and Design,* Upper Saddle River, NJ: Prentice Hall, 1993.

Martin, J., *Systems Engineering Guidebook: A Process for Developing Systems and Products,* Boca Raton, FL: CRC Press, 1997.

Martin, J., and J. Odell, *Object-Oriented Analysis and Design,* Upper Saddle River, NJ: Prentice Hall, 1992.

Rumbaugh, J., et al., *Object-Oriented Modeling and Design,* Upper Saddle River, NJ: Prentice Hall, 1991.

Sage, A., and J. Armstrong, *Introduction to Systems Engineering,* New York: John Wiley & Sons, 2000.

Shlaer, S., and S. Mellor, *Object-Oriented Systems Analysis,* Upper Saddle River, NJ: Prentice Hall, 1988.

Stevens, R., et al., *Systems Engineering: Coping with Complexity,* Hertfordshire, England: Prentice Hall Europe, 1998.

Wheatcraft, L. S., M. J. Ryan, and T. E. Katz, *Needs and Requirements Manual: Needs, Requirements, Verification, Validation Across the Lifecycle,* Hoboken, NJ: John Wiley & Sons, 2024.

Young, R., *Effective Requirements Practices,* Boston, MA: Addison-Wesley, 2001.

Westerman, H., *Systems Engineering Principles and Practice,* Norwood, MA: Artech House, 2001.

[32] For more detail see, for example, Cook, R., D. Malkus, and M. Plesha, *Concepts and Applications of Finite Element Analysis,* New York: John Wiley and Sons, 1989.

[33] Defense Systems Management College (DSMC), *Systems Engineering Management Guide,* Washington, DC: US Government Printing Office, 1990.

[34] Vitech Corporation, *A Primer for Model-Based Systems Engineering,* Second Edition, 2011.

[35] INCOSE, *Systems Engineering Vision 2020,* (INCOSE-TP-2004-004-02,), International Council on Systems Engineering, 2007.

[36] Object Management Group, *OMG Systems Modeling Language (SysML®) Version 2.0,* Needham, MA: Object Management Group, 2025.

[37] Excellent coverage of MBSE and its application can be found in the following texts:

Buede, D. M., and W. D. Miller, *The Engineering Design of Systems: Models and Methods,* New York: John Wiley & Sons, 2006.

Delligatti, L., *SysML Distilled: A Brief Guide to the Systems Modeling Language,* Upper Saddle River, NJ: Pearson Education, 2010.

Dori, D., *Model-Based Systems Engineering with OPM and SysML,* New York: Springer, 2016.

Friedenhal, S., A. Moore, and P. Stein, *A Practical Guide to SysML: The Systems Modelling Language,* Amsterdam: The MK/OMG Press, 2015.

Hetherington, D., *Simple SySML for Beginners: Using Sparx Enterprise Architect,* Austin, TX: Asatte Press, 2019.

Gangl, M., et al., *Modern Methods of Systems Engineering: With an Introduction to Pattern- and Model-Based Methods,* Charleston, SC: Createspace Independent Publishing Platform, 2011.

Holt, J., *Systems Engineering Demystified: A Practitioner's Handbook for Developing Complex Systems Using a Model-Based Approach,* Birmingham, UK: Packt, 2021.

Holt, J., and S. Perry, *SysML for Systems Engineering: A Model-Based Approach,* London, UK: Institution of Engineering and Technology, 2013.

Micouin, P., *Model-Based Systems Engineering: Fundamentals and Methods,* New York: John Wiley & Sons, 2014.

Vitech Corporation, *A Primer for Model-Based Systems Engineering,* Second Edition, 2011.

Weilkiens. T., *Systems Engineering with SysML/UML: Modeling Analysis, Design,* Boston, MA: MK/OMG Press, 2008.

7

Related Disciplines and System Development Approaches

7.1 Introduction

In this last chapter, we discuss two major aspects associated with project management and systems engineering. First, we discuss a number of the other disciplines involved in a project's systems development effort, reflecting on how each is related to systems engineering and project management.

Second, we discuss alternative development approaches. Throughout this book, to best explain the activities and artifacts of systems engineering, we have assumed that life cycle activities are undertaken sequentially. In doing so, we have assumed what is generally referred to as the linear sequential or waterfall approach to system development. There are, however, several other development approaches, each of which has strengths and weaknesses depending on the nature of the system under development. The selection of a suitable development approach is a critical activity early in a system life cycle. This chapter reviews the major development approaches and discusses their applicability.

7.2 Related Disciplines

This section introduces some of the related disciplines that interact with the project management and systems engineering effort during the course of a normal project. We don't attempt to treat any of them in a comprehensive manner since each discipline is a large subject area in itself. Rather, the intention is to show how each of the major disciplines is related to project management and systems engineering.

7.2.1 Business Analysis

A business analyst (BA) plays a crucial role in organizations by bridging the gap between business needs and technological solutions. BAs' principal role is to understand business objectives, analyze data, identify inefficiencies, and recommend improvements to processes, systems, and strategies. Their primary contribution to project management and to systems engineering is to ensure that business requirements are accurately captured and translated into actionable solutions that align with an organization's strategic objectives. In that sense, business analysis contributes most significantly to the preacquisition phase and to the conceptual design activity in that phase where BAs would be responsible for the development of BNR, SNR, and the SyRS.

The most widely recognized guide outlining best practices and standards in business analysis is *Business Analysis Body of Knowledge* [1], published by the International Institute of Business Analysis (IIBA). The IIBA also provides an Agile Extension to the *Business Analysis Body of Knowledge* [2] to assist BAs working in Agile environments by detailing their role in Scrum and other Agile methodologies.

BAs operate across various industries, including finance, healthcare, information technology, government, and retail. They can work within internal business teams, as part of a consulting firm, or alongside project managers, developers, and stakeholders to ensure that solutions deliver maximum value.

BAs perform several key functions within an organization, described as follows:

- *Requirements gathering and analysis:* BAs are responsible for eliciting, recording, and validating business requirements from stakeholders. They use various techniques such as interviews, workshops, surveys, and artifacts analysis to gather and record relevant information.
- *Process improvement and optimization:* By analyzing existing business processes, BAs identify inefficiencies, redundancies, and opportunities for improvement. They use methodologies such as Lean [3], Six Sigma [4], and Business Process Modeling [5] to recommend optimized workflows.
- *Stakeholder management:* Since BAs interact with different levels of an organization, including executives, IT teams, and end users, they act as

a liaison between these groups. They must ensure clear communication and alignment on project goals.

- *Data analysis and decision support:* BAs often analyze data to provide insights into business performance and trends. They may use data visualization tools, reporting software, and statistical models to support decision making.
- *Solution evaluation and implementation support:* Once a solution is proposed, BAs help assess its feasibility, work with technical teams on implementation, and validate that the solution meets business needs. They also play a role in acceptance testing and postimplementation reviews.
- *Records and communication:* BAs create and maintain artifacts such as business needs and requirements, functional specifications, process diagrams, and user stories to ensure clarity and alignment on project objectives.

BAs collaborate closely with project managers and systems engineers to ensure that project objectives continuously align with business needs and provide value. The BA, project manager, and systems engineer all work with stakeholders to define requirements, which the project manager and systems engineer ensure are implemented effectively. In Agile environments, BAs frequently work as Product Owners or Agile Business Analysts, defining user stories and working with development teams to ensure continuous delivery of business value.

7.2.2 Quality Assurance

The term "quality" as used in the context of quality assurance means: "...the totality of feature and characteristics of a product or service that bears on its ability to satisfy stated or implied needs" [6].

With this definition in mind, quality assurance aims to ensure that the products and services required from a system development effort are delivered at the required quality level. This requires planned managerial effort to perform the following:

- Identify the quality assurance objectives and standards;
- Focus on preventative measures rather than cures;
- Collect relevant data and use the data to analyze results;
- Establish and monitor quality performance measures;
- Perform quality audits.

A major role of quality assurance is to ensure that appropriate plans and procedures are in place to perform the design and development tasks. These plans must be well-recorded, approved by the appropriate personnel, and maintained and updated when necessary. It is not, however, satisfactory to merely have the plans and procedures in place. The organization must operate in accordance with those plans and procedures. The quality assurance role will perform audits periodically to ensure that the systems engineering effort is progressing in accordance with the approved formal process. The quality assurance representative requires evidence of the plans and procedures being followed.

7.2.3 Integrated Logistics Support (ILS)

The main aim of ILS is to impact the design of the system to maximize supportability over the entire system life cycle. A definition of logistics support is: "…to assure the effective and economical support of a system throughout its programmed life cycle" [7]. The logisticians will also be interested and involved with maintaining and, where possible, improving system reliability throughout the system's life cycle. The FRACAS process, introduced in Chapter 4, is a systems engineering management issue aimed at doing just that. Both logisticians and systems engineers must combine skills and expertise to make the FRACAS process as effective as possible.

A definition of logistics support is: "…to assure the effective and economical support of a system throughout its programmed life cycle" [8]. To meet this definition, a number of support services are provided (by what is sometimes called the support system) during the utilization phase. The constituent capabilities that provide those support services are acquired during the acquisition phase. They are described as follows:

- *Operating support capability:* This capability encompasses the support elements of operating facilities system operators, support equipment, operator manuals and technical data, operating support procedures, and operating support information systems.
- *Engineering support capability:* Engineering support encompasses the support elements of engineering facilities, engineering personnel, engineering support and test equipment, engineering technical data, engineering processes, the engineering information management system, and software support.
- *Maintenance support capability:* This capability includes the support elements necessary to develop, establish, and integrate a maintenance support system capable of sustaining a system throughout its life. The capability comprises the elements of maintenance facilities, maintenance

personnel, maintenance support and test equipment, maintenance technical data, maintenance processes, and the maintenance information management system.

- *Supply support capability:* Supply support encompasses the support elements of supply facilities, supply personnel, supply support equipment, supply technical data, supply processes, supply information management system, spares, and packaging.
- *Training support capability:* The training support capability includes the support elements of training facilities, training personnel, training equipment, training materials and other technical data, training processes, and the training information management system.

Although the support capabilities have been dealt with independently in this section, it must be stressed that the elements are strongly interrelated. It is the role of the logistician to ensure that the elements are considered during the systems engineering processes and throughout the system life cycle.

7.2.4 Operations

Operations is the generic name sometimes given to the actions of the users (see Section 1.6) and support personnel who will operate, support, and maintain the system when it is in service. Operations personnel are vitally important to the project, and they play differing roles as the project progresses through the acquisition phase. For example, in conceptual design and preliminary design, the operators are likely to be heavily involved in the correct and complete definition (and interpretation) of the stakeholder requirements. This is vital because these interpretations will flow through the entire systems engineering effort and result in specific system performance. During the latter stages of the acquisition phase, the operators will be involved in various testing activities. Depending on the nature of the system under development, aspects of the system's operation (and therefore testing) will only be possible using appropriately trained personnel. Consider for example the testing of an attack helicopter's control system or the operation of an air defense radar system capable of tracking 150 targets simultaneously. While this can be verified by the supplier during AT&E, the acquirer will want to validate the requirement when the trained pilot is on board.

7.2.5 Software Engineering

Software engineering is defined as: "...the technological and managerial discipline concerned with systematic production and maintenance of software products that developed and modified on time and within cost estimates" [9]. Software engineering has a set of processes, management requirements, tools, and related disciplines that are very similar to systems engineering.

ISO/IEC/IEEE 12207 [10], first published in 1995, is the primary international standard that defines the processes required for developing and maintaining software. On May 27, 1998, ISO/IEC/IEEE 12207 officially replaced MIL-STD-498 for the development of US Department of Defense software systems. The 2017 publication of ISO/IEC/IEEE 12207 has been harmonized with ISO/IEC/IEEE 15288 and adopts the ISO/IEC/IEEE 15288 process model with one slight change in the name of one process—the "System Requirements Definition" process is renamed the "System/Software Requirements Definition" process.

The *Software Engineering Body of Knowledge (SWEBOK)* defines eighteen knowledge areas [11], described as follows:

- *Software requirements:* Introduces the requirements engineering process including the subareas of requirements engineering process, requirements elicitation and requirements analysis, software requirements specifications, requirements validation, requirements management, and requirements tools.
- *Software architecture:* Introduces software architecture fundamentals, description, process, and evaluation.
- *Software design:* Covers the process of defining the architecture, components, interfaces, and other characteristics of the software.
- *Software construction:* The underlying software engineering principles by which software is crafted.
- *Software testing:* The process through which defects and problems are identified and software quality is improved.
- *Software engineering operations:* Introduces software engineering operations fundamentals, planning, delivery, and control.
- *Software maintenance:* Sustains software products throughout the life cycle to ensure that problems are identified and resolved, training is conducted, and modifications are implemented in a controlled manner.
- *Software configuration management:* Introduces the configuration management of software in a manner similar to that described in Chapter 5.
- *Software engineering management:* Describes the organizational, project management, and measurement aspects of software engineering.
- *Software engineering process:* Encompasses the technical and managerial processes performed during acquisition, maintenance, and disposal of

software products, as well as the definition, implementation, measurement, management, change, and improvement of the software engineering process itself.

- *Software engineering models and methods:* Defines, in a manner similar to that used in this text, the tools, models, and methods associated with the software engineering process and management. The SWEBOK includes both the software development environments and software development methods.
- *Software quality:* Deals with software quality assurance and verification and validation.
- *Software security:* Introduces software security fundamentals, management and organization, engineering and processes, security engineering for software systems, software security tools, and domain-specific software security.
- *Software engineering professional practice:* Covers the knowledge, skills, and attitudes that software engineers need to practice software engineering in a professional, responsible, and ethical manner.
- *Software engineering economics:* Covers the process of making decisions regarding software engineering in a business context.
- *Computing foundations:* Covers the development and operational environment (the computer and its various components) in which software evolves and executes.
- *Mathematical foundations:* Helps software engineers comprehend well-understood, unambiguous logic upon which software is based, which in turn is translated into programming language code.
- *Engineering foundations:* Covers the engineering foundational skills and techniques that are useful for a software engineer.

Software products pass through a development process that is tied into the system life cycle leading up to the identification of CIs and the described HWCIs and CSCIs. Each CI is managed individually and follows an individual development cycle.

Software engineering has some unique challenges that must be addressed jointly by the software and systems engineering specialists. Software is inherently difficult to test due to the number of variables often associated with its successful operation. Measurement of software reliability is also difficult. Software maintenance and support are often overlooked as many assume that software does not require maintenance and support once fielded. Software integration with host hardware brings about its own set of challenges as does the issue of firmware as a specialized category of software. These challenges mean that traditional approaches to design, development, integration, testing, and support derived from hardware-dominated environments are not always satisfactory in a software environment. To that end, software engineering is recognized as a specialty engineering discipline in this text, and systems engineers should treat it accordingly.

7.2.6 Hardware Engineering

Hardware engineering processes proceed in a similar manner to those of software engineering following the selection of the HWCIs. The HWCIs are progressively broken into components and units in a similar way to the CSCs and CSUs.

Hardware units are designed and developed and then tested via unit-level testing. Once successful, the units are integrated and retested as they form the hardware components. This process is repeated to form and test the HWCIs. Finally, the HWCIs and CSCIs are combined, and system-level testing is conducted.

Hardware and software engineers must work closely to ensure that interfacing issues are identified and addressed. Interfacing issues are sometimes addressed via an ICWG, which meets regularly during the system development process.

As with software engineering, recording and management of the development process will be managed on an individual HWCI basis. This allows different approaches to be used for each HWCI. For example, one HWCI may make use of OTS equipment with minor modifications while another HWCI may require a bespoke design and development effort.

7.3 System Development Methodologies

While waterfall development is often used successfully in system development, the main reason it is used in this text is that it provides a logical, sequential flow of activities and deliverables that support the explanation of systems engineering processes and activities. Additionally, waterfall development is generally considered to be the basic building block upon which the alternative approaches such as incremental, evolutionary, and spiral development are built [12, 13]. Furthermore, in a 2004 survey [14], approximately one-third of software development projects reported using a waterfall development approach. A solid understanding of waterfall development is therefore very useful, as it is still valid in many circumstances.

Like most aspects of systems engineering, it is simply not possible (nor desirable) to state categorically that a given approach to development is either universally acceptable or universally useless. In fact, the urge to develop a universally applicable model while dismissing other approaches promotes a "one size fits all" mindset, which can lead to an inappropriate "cookbook" approach to systems engineering. There are simply too many variables in play in today's complex project environment. Projects, by definition, are undertakings to produce unique outputs [15] and generally require a bespoke systems engineering and acquisition approach to be applied. Successful systems engineering relies on intelligent, well-informed, and experienced professionals determining the most appropriate way to execute a technical program.

To help this decision-making process, this section investigates the characteristics of the principal approaches to system development. We examine waterfall development, incremental development, evolutionary development, spiral development, and Agile methodologies and describe when these alternatives may be considered a more pragmatic approach to development and acquisition.

7.3.1 Waterfall Development

Waterfall development, a variation of which is illustrated in Figure 7.1, was established as a response to the emerging software engineering discipline [16] but has since been adopted by the wider engineering community. The approach gets its name from the fact that it relies on the development of a complete set of requirements at the system level, which influences and cascades into the subsystem requirements, which in turn, drives the requirements for the assemblies and components. At each transition, a baseline tends to be established—commonly, the FBL, the ABL, and the PBL. This idea gives rise to the visualization of requirements cascading down a waterfall as shown in Figure 7.1. Because of its nature, waterfall development is also often called the linear sequential model.

While the flow of a waterfall model is in the forward direction, there is a recognition that a mechanism is needed by which experience at the lower levels can be used to revisit the activities at the higher level. Being able to manage changing requirements that are driven by discoveries in later systems engineering stages is a critical element of a pragmatic systems engineering process. While the waterfall model does allow this feedback, it is often difficult to implement and limited to feeding back only to the preceding stage if the artifacts produced by that stage require correction. Consequently, the waterfall model is often criticized as being largely a one-shot process.

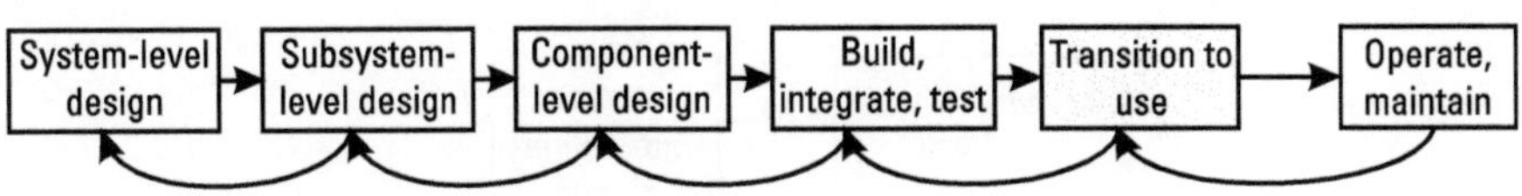

Figure 7.1 Cascading activities in the waterfall approach.

So, despite the accommodation of limited feedback, the waterfall approach to development is most effective when a complete and accurate set of requirements exists at the system level and when those requirements remain relatively stable throughout the subsequent acquisition process. This leads to the approach sometimes being referred to, rather disparagingly, as a grand design approach.

The classic waterfall development approach has been adapted into an equally well-known construct called the Vee diagram, which was first presented to the systems engineering community in 1991 [17]. The Vee diagram, as used in this text, applies to taking a system from conceptual design through to the end of detailed design and development and uses the concepts pioneered by the waterfall approach but arranged in the shape of a Vee as illustrated in Figure 7.2, in order to reflect the systems engineering philosophy of top-down design with bottom-up development.

Whether shown in the cascading construct of Figure 7.1, or in the shape of the Vee in Figure 7.2, waterfall development is characterized by an effort to understand and record the complete scope of the system before undertaking the subsequent engineering design and acquisition effort.

In some instances, the use of the waterfall model is appropriate—if we are building a bridge across a wide channel, for example, we cannot start building until we have established all requirements, completed all detailed design, and so on. Additionally, waterfall development is almost universally applied in fixed-price contracting, which requires the development of a complete WBS (that can then be scheduled and costed) before the contract can be signed.

In many cases, however, it is not possible to establish requirements that remain stable over the system life cycle. A large and complex system development requiring integration with other (external and changing) complex systems is a situation in which establishing accurate and stable requirements upfront

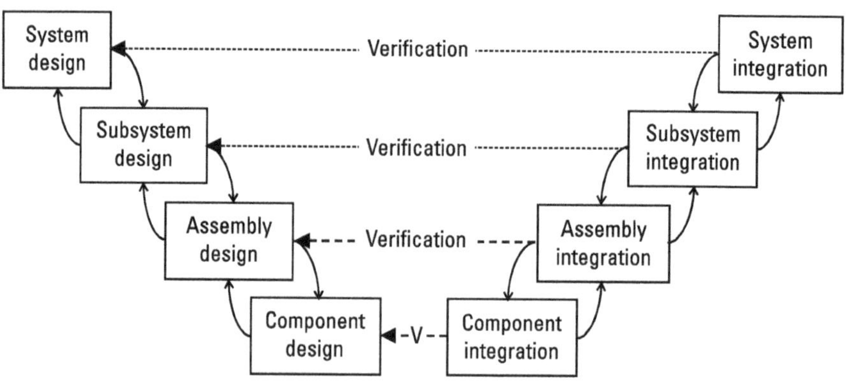

Figure 7.2 Vee representation of the waterfall approach.

may be difficult (and may, in fact, be a liability). An accurate and complete set of requirements is also unlikely when the capability being introduced is completely new to the stakeholders, particularly when it is to be a unique system. In these cases, the stakeholders simply may not be able to articulate accurately their requirements.

Difficulties associated with developing a comprehensive set of system-level requirements upfront have been well-recorded in many other places. Yet, weaknesses in the requirements management process are a major cause of project failure—we know that if we do not get the requirements right up front, there is a significant risk of project failure.

In addition to the requirements challenges, the acquisition process is generally related to some form of contract, binding the acquirer and the supplier of the system. The waterfall approach relies on sufficient money and time being available to complete the entire development effort in a single iteration. As with requirements, it is not always possible to commit the necessary time or money to large, single development efforts.

It is not only system-level requirements that tend to change over time. A characteristic of modern technology is the pace with which it changes. As the waterfall approach is based on the establishment of progressively more detailed baselines, it may be difficult in large projects to take advantage of opportunities presented by advances in technology. Design decisions and approaches tend to be made early in waterfall development, which may make design change prohibitive. Other areas of change on typical system developments that may cause problems with the waterfall approach include changes in any of the external systems with which the system under development must be integrated; changes in the operating or support environments; and changes to any of the constraints under which the system must be designed, produced, operated, or supported.

We must conclude by noting, however, that waterfall development, which places strong emphasis on early elicitation of requirements, is very useful in the following situations:

- When the requirement set is small and well-understood;
- When it is used as the basis of smaller builds in other development methods;
- When fixed-price contracting is used.

7.3.2 Incremental Development

In many cases, there may be valid reasons why the complete project scope cannot be developed and delivered in a single iteration, such as the following.

- *Limited capability early:* Sometimes, some degree of capability is required by the users as soon as possible, without having to wait for the full development time.
- *Insufficient funds:* The project budget is an example of a project constraint that needs to be considered early in any system development. A smaller than required project budget may preclude a classic waterfall approach (addressing the entire scope of the system) from being employed.
- *Insufficient time:* In some circumstances, stakeholders may require a minimal set of functionalities to be delivered urgently, meaning that there is insufficient time to design, construct, and deliver the entire project.
- *Incomplete understanding of requirements:* It is not possible to completely bound the scope of requirements early in the program. Consequently, a classic waterfall approach becomes either a gamble (based on a best-guess scope) or a program plagued by continual change as the understanding of scope or functionality matures.
- *Risk management:* Perhaps all the above issues could be couched as risk-management issues. Incremental development assists in the management of risk by choosing small, well-defined, manageable work packages.

Incremental development starts with a good understanding of scope and level of functionality desired from the system and allocates different aspects of the system scope to a series of builds or increments that will be designed and delivered over time. Normally (as illustrated in Figure 7.3), the first increment introduces a kernel of basic functionality, which is then augmented over subsequent builds with progressive enhancements to scope until the entire system functionality is delivered [18].

Incremental development tends to deliver partial capability more quickly than if a traditional waterfall approach had been adopted. This reduces the likelihood of changes occurring to requirements, constraints, external systems, or the environment during an increment of the development.

Each incremental build may be executed using waterfall development where the top-level requirements of the first build drive the design of that build. If users would like the opportunity to develop or clarify their requirements between builds, the first build may be fielded for a period of use and exploration in order to collect data to inform subsequent builds. The top-level requirements of subsequent builds are refined to take account of the feedback in order to improve on the first build. In this way, it is possible to consider incremental development as a series of smaller waterfall developments, as illustrated in Figure 7.4.

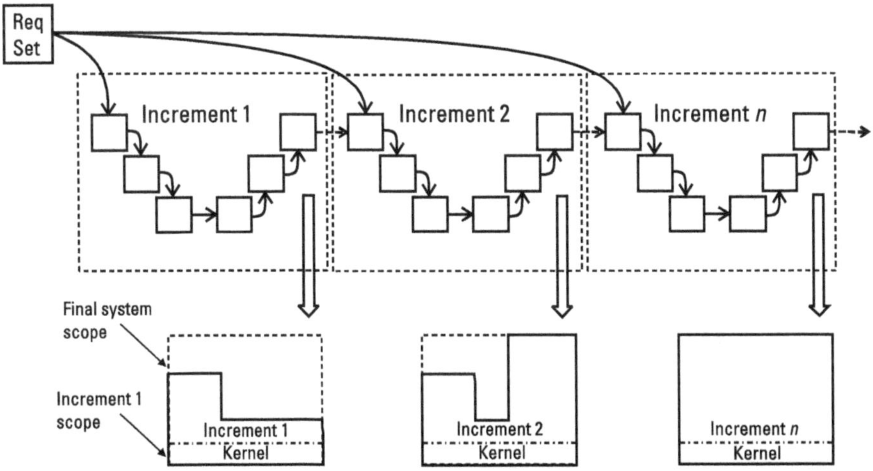

Figure 7.3 Incremental development building on system functionality.

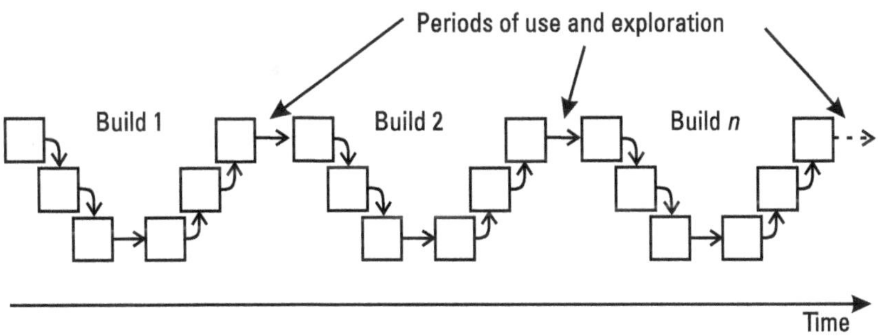

Figure 7.4 A series of waterfall builds.

The builds (and therefore the waterfalls) may be separated by considerable time (as illustrated in Figure 7.4) or may overlap to some degree. Overlapping builds may occur in instances where requirements sets are frozen, or baselined, to support an initial build but continue to mature after they have been frozen. Requirements that change after the initial baseline has been frozen can be collected and fed into the baseline of the second build, and so on. This may give rise to a degree of overlapping as illustrated in Figure 7.5 where successive requirements baselines (RBLs) are shown as a function of time. Either way, the establishment of a number of builds that are clearly defined at the outset not only allows the use of the waterfall approach for each build, but it subsequently allows fixed-price contracting for each build.

If incremental development is being considered, special consideration should be given to including requirements in the initial build that address long-term design goals such as expandability and upgradability. This will help ensure

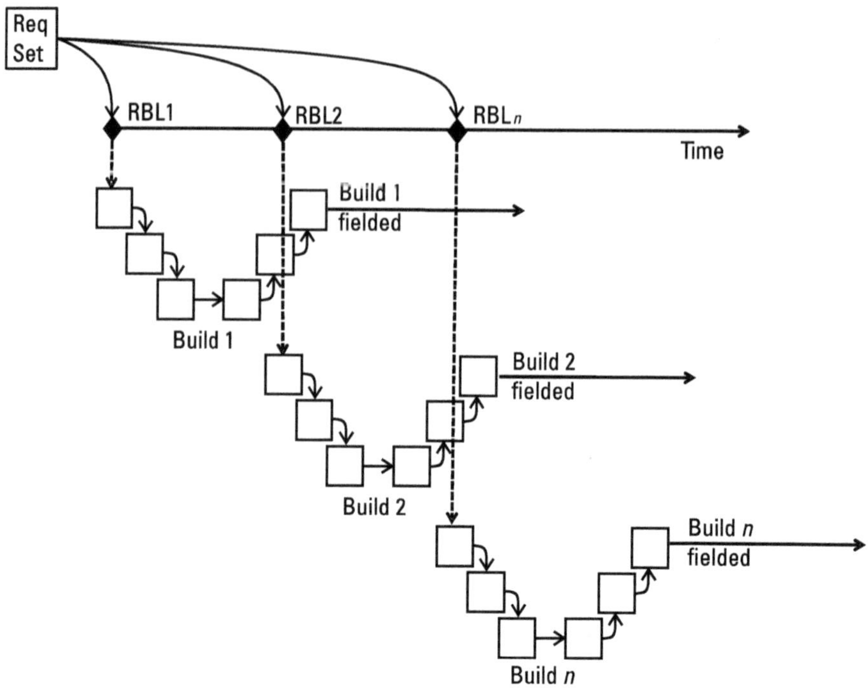

Figure 7.5 Overlapping incremental builds.

that subsequent builds are able to maximize use of the initial build without having to replace or upgrade significant percentages of the system each time. For example, if the system being developed is a combination of hardware and software, consideration should be given to provision of spare computer capacity; spare communication bandwidth; and spare mass, space, and power allowances in the initial build. Even when expandability and upgradability is properly considered in the initial build, incremental development strategies must recognize that design decisions made during the initial build will constrain the design of subsequent builds.

The contractual mechanism to be used to bind the acquirer and developer needs to be considered carefully when employing incremental development. While the scope, time frame, and costs associated with the initial build may be well-understood and easy to translate into a contract, this may not be the case of the subsequent builds. Both the acquirer and the developer need to be interested in how to broker a contractual agreement encapsulating incremental development.

Finally, this text has emphasized the importance of stakeholder engagement in the systems engineering process, including the need to manage stakeholder expectations. With incremental development, management of stakeholder expectations becomes critical. The stakeholders must understand and agree on the level of functionality being delivered in each of the planned increments. This guards against stakeholder disappointment or even rejection when initial builds fail to deliver the entire scope that is ultimately being aimed for. The aim of the incremental development is to eventually deliver complete functionality progressively as illustrated in Figures 7.3 and 7.4.

7.3.3 Evolutionary Development

Evolutionary development is very similar to incremental development [19] in that it aims to arrive at the desired level functionality and scope over a period of time using a number of design iterations. Although incremental development and evolutionary development are very similar in their aims, the defining difference is that incremental development is a series of discrete builds with deliberate augmentation of functionality, whereas evolutionary development is a more continuous evolution toward the final system scope including the development of some functionality that is later discarded [20]. Additionally, the presumption in evolutionary development is that, not only are the requirements not completely known in advance, but that the development process is used to elicit unforeseen requirements and to identify weaknesses in early sets.

The two approaches have different applications, because stakeholders sometimes have little idea about what they want from the solution at the outset of the project but are able to provide meaningful critique of proposed solutions when they see them. In these situations, the stakeholders are not able to describe the complete set of requirements for the system they need or agree on how that will be delivered as a series of incremental builds. An evolutionary approach may be able to assist. In other situations, stakeholders are keen for some level of solution to at least help them with their problems and acknowledge that they are willing to accept an imperfect initial solution.

The initial stage of the evolutionary development, therefore, focuses on delivering something of use, mindful of the fact that significant variation to functionality is allowed until the series of builds iterates toward the final system scope. This is illustrated in Figure 7.6. As with incremental development, executing an evolutionary technical program may be thought of as a progression of separated by some period of time or, as illustrated in Figure 7.7, evolutionary builds may be overlapped in time in order to speed up the delivery of functionality to stakeholders.

Given the similarity in the concepts of both incremental and evolutionary development, the same set of considerations discussed with incremental development need to be applied to evolutionary developments. Stakeholders must be involved and appraised of the strategy, and the constraining nature of design decision made in early evolutions needs to be carefully considered.

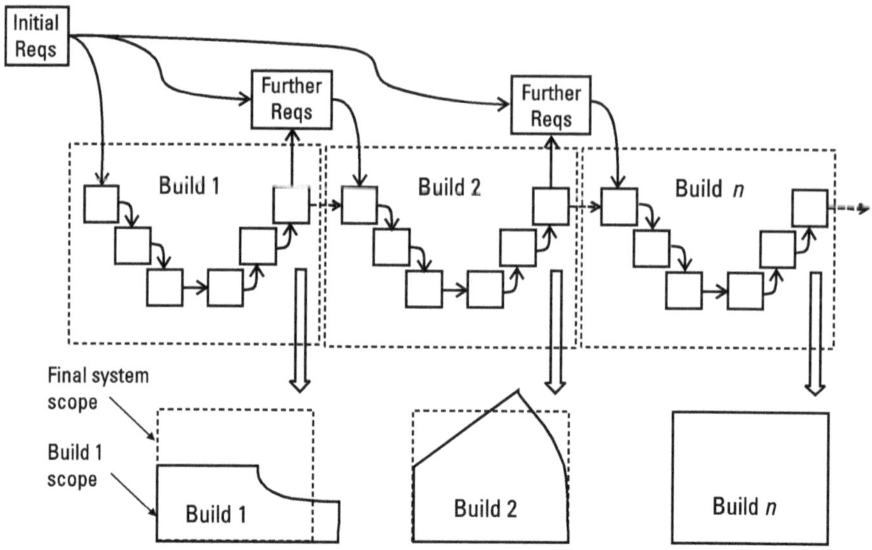

Figure 7.6 Evolutionary development of functionality.

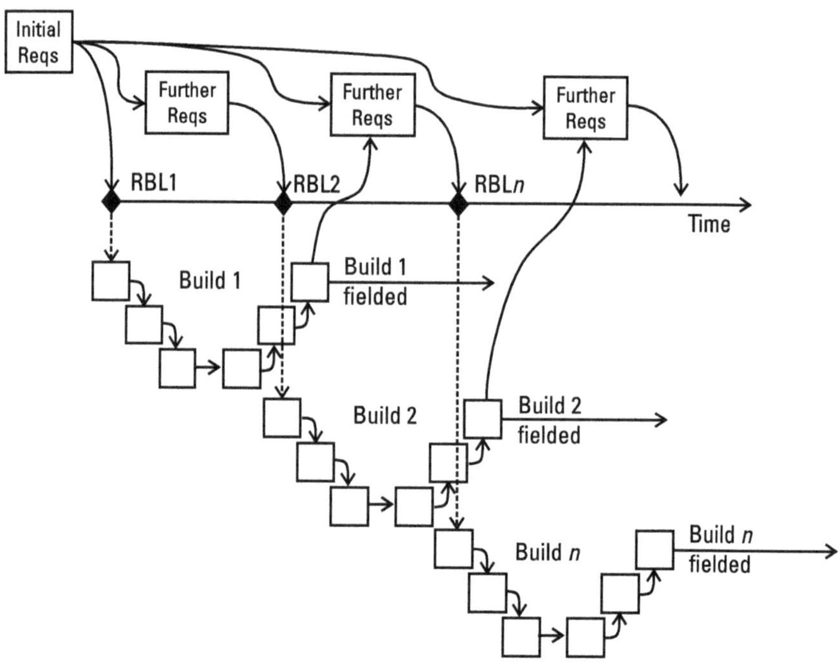

Figure 7.7 Overlapping evolutionary builds.

7.3.4 Prototyping and System Development

Prototyping is a pragmatic way of exploring different aspects of system concepts and requirements before making final decisions and committing to a particular acquisition path. Prototyping continues to be an important part of many design and development efforts as they proceed from system to subsystem, to component-level design.

At a system level, a prototype may be produced to provide some limited functionality to key groups of stakeholders in order to facilitate exploration of their requirements for the mature system.

Prototyping is not the only way in which to gain a better understanding of the capability requirements before embarking on an acquisition. Another popular tool or activity is called concept exploration and demonstration. In this case, stakeholders can access or experience systems that deliver capability that is similar in scope or functionality to the capability they desire. These similar systems are not prototypes because they are generally already in-service with other companies or organizations. The stakeholders observe and, in some cases, use these systems to explore and mature their understanding of their own requirements for the capability. Acquisition is then able to proceed from a basis of knowledge rather than one of guesswork.

In this way, prototyping and concept exploration and demonstration assist in the early requirements analysis task. A range of other tools can also be used to help with requirements analysis but are not discussed specifically here, including scale models, mockups, functional models, and simulations.

The key difference between a prototype and the early builds in either incremental or evolutionary developments is that a prototype is provided with a minimum of formal process simply to assist the stakeholders in exploring their requirements. The requirements from the prototyping phase are then used as the key input in a development phase that employs a more formal methodology such as waterfall, incremental, or evolutionary acquisition. The prototype does not generally go on to become a fielded system but may continue to play a role in the requirements determination process, especially if incremental or evolutionary methodologies are employed.

Another difference between a prototyping process and early incremental or evolutionary builds is that a prototype is used to explore what the stakeholders do not know about their requirements, whereas early incremental or evolutionary builds aim to deliver what the stakeholders do know about their requirements.

Developing the prototype—although a less formal process than a complete system development—should be done using a basic process of some sort. After all, even prototypes can cost significant sums of money, and their acquisition and use needs to be controlled. The process may as simple as determining the need for the prototype by agreeing on its purpose, designing/building or acquiring the prototype, and then using and refining it until the prototyping phase has been completed. This basic process is illustrated in Figure 7.8.

In some cases, the prototype may eventually develop into the final system via an incremental or evolutionary development process. This strategy is sometimes called evolutionary prototyping [21]. If this is not the strategy, the prototype should be disposed of when no longer required to ensure that is not misused.

The use of prototyping as an exploration tool does not end at the system level illustrated in Figure 7.8. Prototyping is an important tool or technique that gets used at all levels of the system development to discover and confirm requirements and constraints. For example, designing and agreeing on user interfaces during the preliminary design or detailed design and development phases commonly uses prototyping and key stakeholders to confirm the design adequacy of the proposed solutions before finalizing the design artifacts.

7.3.5 Spiral Development Model

Boehm [22] pioneered the spiral model, illustrated in Figure 7.9, as a software development and enhancement framework. It is interpreted and presented here from a systems engineering perspective, rather than from a software perspective.

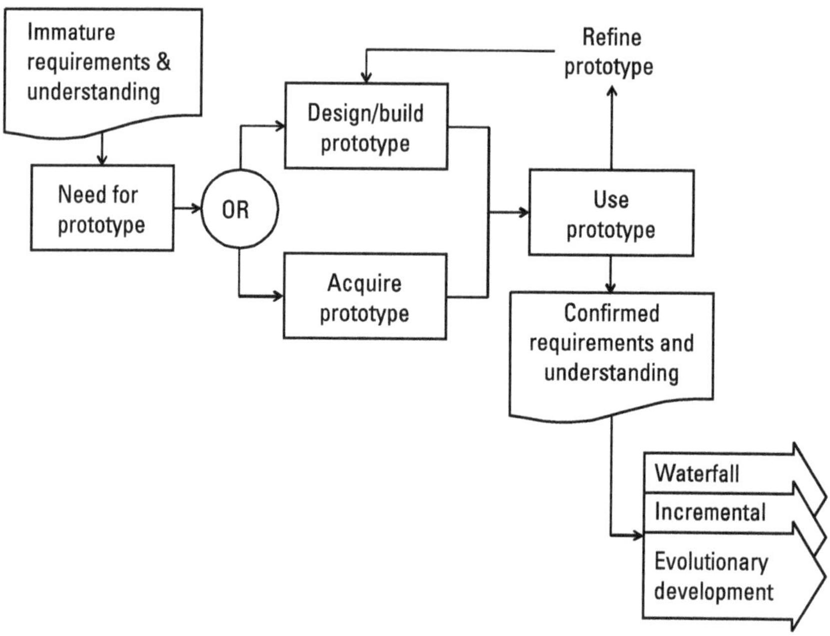

Figure 7.8 Prototyping preceding design and development.

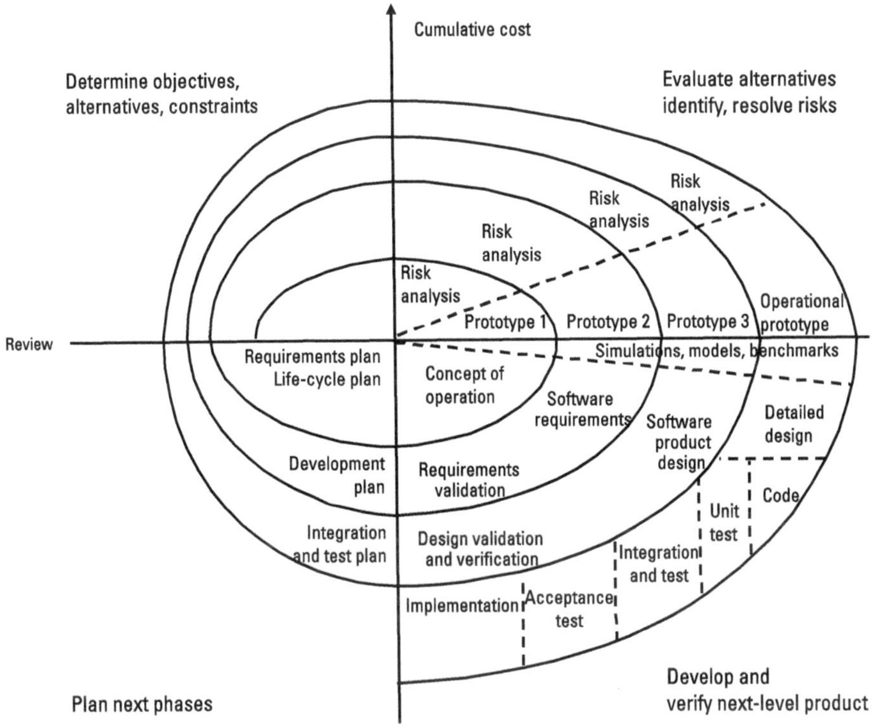

Figure 7.9 Spiral development model. (*After:* [22].)

The spiral model is not a development methodology per se but a process by which the most appropriate development methodology can be selected for each system development effort. The spiral model does not presume any particular methodology at the start of the process, nor does it require the same methodology to be adopted for each cycle of the spiral. Spiral development recognizes that things change over the course of a development, possibly necessitating strategy adjustment or change, and supports adoption of the most appropriate process model such as waterfall, incremental, or evolutionary depending on circumstances. For example, if aspects of the system development have many unresolved risks and poorly defined user requirements, evolutionary development augmented by prototyping might be employed. If the risks in other aspects of the system development are lower, and the requirements are well-defined and stable, a more traditional waterfall approach may be adopted. The spiral development model is therefore best viewed as a meta-model [23] that generates other models.

Each cycle of the model follows the same sequence of activities, described as follows:

- *Determine objectives, alternatives, and constraints (top left of Figure 7.9):* The key requirements of the system element under consideration are determined. The alternatives for achieving these requirements are identified. This may include options such as reuse of existing designs, use of OTS technology, or the need to proceed with some form of design and development. The constraints that are relevant to the system element are also identified and may include cost and time limitations, external systems and interfaces, environmental issues, and laws/legislation.
- *Evaluate alternatives and identify and resolve risks (top right of Figure 7.9):* The alternatives are considered in light of the requirements and the constraints. Each alternative will typically come with a different set of risks. The spiral model incorporates prototyping, modeling, and simulation as examples of tools that can help consider those risks and determine risk management strategies.
- *Develop and verify next-level product (bottom right of Figure 7.9):* The next level of product is then produced based on displacement from the center of the spiral. Close to the center of the spiral (the early stages of the development), the products might be subsets of system-level requirements. Further out are the sets of subsystem requirements, and further still are the detailed design and integration products. Regardless of product, the spiral model includes relevant validation and verification activities before moving onto the next stage.
- *Plan next phases (bottom left of Figure 7.9):* In this phase the verified and validated products from the previous phase are used to plan for the next iteration of the spiral. Early on, the plans will be project- and system-level management plans but later on, more detailed plans such as integration and test plans will be developed.
- *Review by stakeholders:* Each cycle of the spiral concludes with a period of review by relevant stakeholders (shown along the x-axis in Figure 7.9). This period of review allows the stakeholder to review the products and decisions made to this point and the plans for the subsequent phases. The reviews facilitate decision making and renewed commitment to projects by focusing on risk and encouraging early and ongoing decisions about whether a system development should continue or be canceled.

Spiral development, with its flexibility to make ongoing decisions and make changes to the direction of system developments, has been found to be effective in situations where organizations are undertaking internal system developments. There are some challenges to be addressed when extending the model to commercial developments where an acquirer and a developer are bound by contractual terms and conditions relating to the development. In these cases, traditional contracting mechanisms such as fixed-price contracting may not sufficiently accommodate the changes to the development process encouraged under spiral development. Spiral developments may require some form of cost-plus/risk-sharing contracting mechanisms to be employed. This type of contracting mechanism is not universally popular with some acquirer organizations, which may rule out using spiral development model.

7.3.6 Agile Development

Agile development refers to a range of approaches to software development that aim to deliver value faster, with improved quality and greater ability to respond to changes throughout the life cycle. Agile development therefore encompasses a number of development methodologies in which the approach to development is iterative/evolutionary, with adaptive planning, allowing requirements and solutions to evolve through closer collaboration of self-organizing, cross-functional teams and acquirers. Agile is a mindset defined in the Agile Manifesto [24] with four values, guided by 12 principals, and manifested through many different practices [25]. Although originally focused on software development, Agile approaches are now commonly considered to be applicable to all aspects of system development. It is worth noting that while the term "agile" became popularized after the Manifesto, the approaches and techniques being used by project teams today existed before the Agile Manifesto—by many years and, in some cases, decades.

Scrum, the most widely used Agile methodology, describes a process framework comprising short, time-boxed development cycles, called sprints, that have light documentation overhead with an emphasis on delivering early functionality and being agile enough to accommodate rapidly changing business environments—sprints are typically from one to four weeks long. Scrum involves three main roles: ScrumMaster, Product Owner, and Developers (collectively, the Scrum Team). Functionality (feature requests) are described in terms of a User Story (a templated form of use case).

Shorter iterations, or sprints, aim to minimize the overall risk in development by ending each iteration with a working product that, through multiple iterations, will allow the product to adapt to changes and to reflect the stakeholders' increased understanding of the functions required of it. Development teams are normally small and colocated to facilitate efficient face-to-face communication and reduce the time required for each development cycle. The development team includes an acquirer representative, and each iteration ends with their review of progress and reevaluation of priorities.

Agile practices can be applied to each activity in the system life cycle and in most cases can be applied to one specific activity. For example, there is an

Agile Extension to the Business Analysis Body of Knowledge [26] that provides BAs with seven key guidelines for the practice of business analysis within an Agile environment, as well as specific techniques that allow Agile teams to deliver value. In project management, the PMI, in conjunction with the Agile Alliance provides an Agile practice guide that is an adjunct to the *Project Management Body of Knowledge* to provide tools, situational guidelines, and an understanding of the various Agile practices to enable better project outcomes [27].

7.3.7 Contracting Approach Versus Development Approach

In some cases, such as in-house developments, the relationship between the acquirer and the supplier (the tasking activity and the performing activity) may be such that the development approach and the contracting approach align (that is, the waterfall contracting approach aligns with a waterfall development approach, evolutionary with evolutionary, Agile with Agile, and so on). However, this does not necessarily have to be the case and often may not be in the larger government contracts.

Consider the case in which the acquirer and developer contract may be implemented using an evolutionary approach where, as shown in Figure 7.10, the requirements are collected and analyzed, and a subset is chosen to be formally contracted with a developer—the contract is therefore based on general conditions (covering such factors as skill sets, daily rates, deliverable formats, and verification arrangements), and each build is then an amendment to that contact (in some form of deed of variation) that identifies the next subset of requirements, the associated SOW, and budget and schedule. The developer develops that build, which is verified at an FQR, validated, and fielded. Feedback from that build is used to inform the pool of potential requirements (adding, deleting, and modifying requirements), from which the scope of the next build is developed, and an appropriate contract variation is negotiated for that next subset of requirements. However, as shown in Figure 7.11, in that build the developer may apply any system development approach—Agile for example, because it suits the nature of the build. In the next build, the developer may choose a linear sequential approach, and so on.

7.4 Summary

To facilitate our explanation of the systems engineering process throughout this book, we have used a waterfall approach to systems development, without discussing in great detail the development and acquisition context within which it might be undertaken. As we have noted in this final chapter, however, there are a number of other development approaches, each of which has strengths and weaknesses depending on the nature of the system under development. The

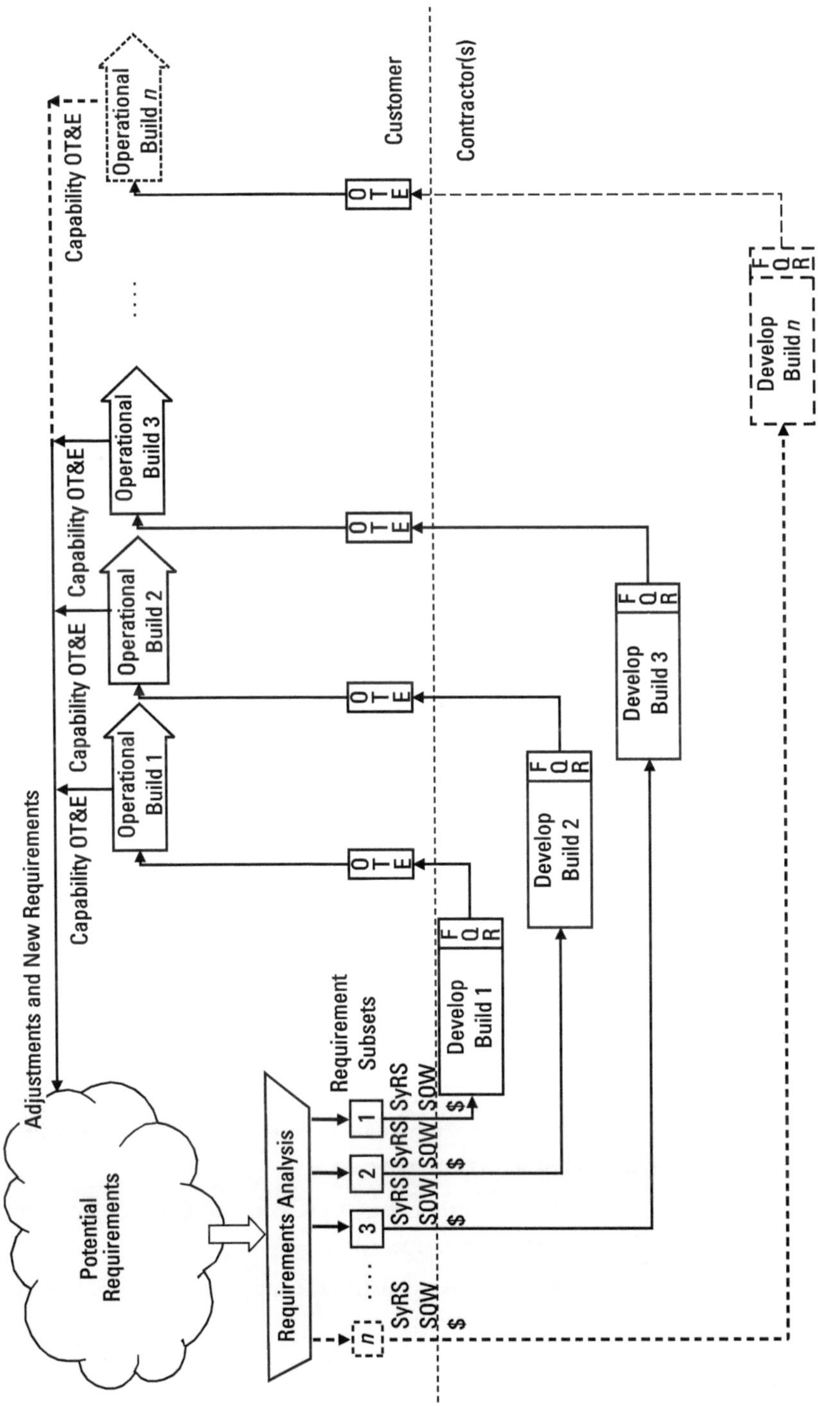

Figure 7.10 Contracting by build in an evolutionary system development.

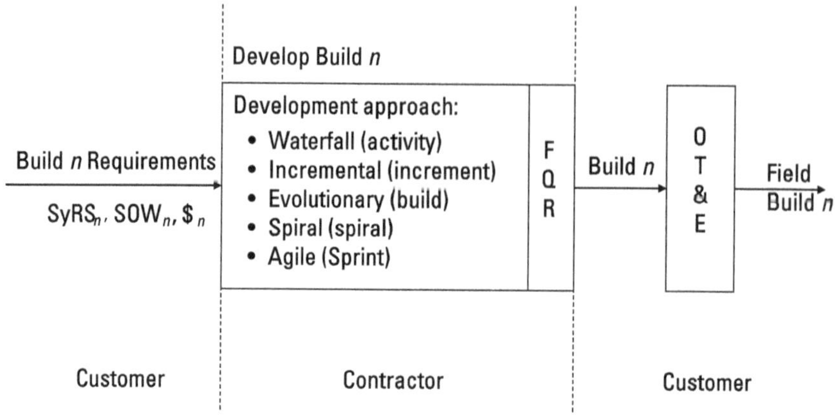

Figure 7.11 Development of a build does not necessarily need to employ the same development methodology as the contracting approach, nor does each build need to use the same development methodology.

selection of a suitable development approach is an essential activity early in a system life cycle—from a project management perspective, it is a critical early decision since it drives the activities and their schedule.

While the linear-sequential waterfall development approach is often used successfully (particularly when used in conjunction with fixed-price contracting), the main reason it is used in this text is that it provides a logical, sequential flow of activities and deliverables that support the explanation of systems engineering. A solid understanding of waterfall development is also required since the waterfall approach can be considered as the basic building block upon which the alternative approaches such as incremental, evolutionary, spiral, and Agile development are built.

Review Questions

1. Explain the role of business analysis in a project, and describe how the discipline relates to systems engineering.
2. Explain the role of quality assurance in a project, and describe how the discipline relates to systems engineering.
3. Explain the role of logistic support in a project, and describe how the discipline relates to systems engineering.
4. Briefly describe each of the five constituent support capabilities of logistics support.
5. The *Software Engineering Body of Knowledge* details 18 knowledge areas that are applicable to that discipline. Briefly describe each of these

knowledge areas and explain how each relates to the discipline of systems engineering.

6. Briefly explain (with the aid of a diagram) the waterfall approach to system development and under what circumstances the use of the approach is appropriate/inappropriate.
7. Briefly explain (with the aid of a diagram) the incremental approach to system development and under what circumstances the use of the approach is appropriate/inappropriate.
8. Briefly explain (with the aid of a diagram) the evolutionary approach to system development and under what circumstances the use of the approach is appropriate/inappropriate.
9. Briefly explain (with the aid of a diagram) the spiral approach to system development and under what circumstances the use of the approach is appropriate/inappropriate.
10. Briefly explain (with the aid of a diagram) the Agile approach to system development and under what circumstances the use of the approach is appropriate/inappropriate.

Endnotes

[1] International Institute of Business Analysis (IIBA), *A Guide to the Business Analysis Body of Knowledge (BABOK Guide)—Version 3,* International Institute of Business Analysis, 2015.

[2] International Institute of Business Analysis (IIBA), *Agile Alliance, and Project Management Institute (PMI), Agile Extension to the BABOK Guide—Version 2,* International Institute of Business Analysis, 2017.

[3] Womack, J. P., and D. T. Jones, *Lean Thinking: Banish Waste and Create Wealth in Your Corporation,* New York: Simon & Schuster, 1996.

[4] Pande, P. S., R. P. Neuman, and R. R. Cavanagh, *The Six Sigma Way: How to Maximize the Impact of Your Change and Improvement Efforts,* New York: McGraw Hill, 2000.

[5] Dumas, M., et al., *Fundamentals of Business Process Management,* Berlin: Springer, 2018.

[6] Kerzner, H., *Project Management—A Systems Approach to Planning, Scheduling, and Controlling,* New York, NY: Van Nostrand Reinhold, 1995, p. 1041.

[7] Blanchard, B., *Logistics Engineering and Management,* Upper Saddle River, NJ: Prentice Hall, 1992, p. 11.

[8] Blanchard, B., *Logistics Engineering and Management,* Upper Saddle River, NJ: Prentice Hall, 2003.

[9] ISO/IEC/IEEE 12207, *Systems and Software Engineering—Software Life Cycle Processes,* 2017.

[10] Fairley, R., *Software Engineering Concepts,* New York: McGraw-Hill, 1985.

[11] Washizaki, H. (ed.), *Guide to the Software Engineering Body of Knowledge (SWEBOK Guide), Version 4.0,* IEEE Computer Society, 2024, www.swebok.org.

[12] Forsberg, K., and H. Mooz, *Systems Engineering Overview,* Cupertino, CA: Center for Systems Management, 1996.

[13] Forsberg, K., and H. Mooz, "Application of the 'VEE' to Incremental and Evolutionary Development," *Proceedings of the NCOSE Conference,* St. Louis, MO, July 1995.

[14] Neill, C. J., and P. A. Lapante, "Requirements Engineering: The State of the Practice," *IEEE Software,* Vol. 20, No. 6, December 2003.

[15] Project Management Institute, *A Guide to the Project Management Body of Knowledge,* Upper Darby, PA: Project Management Institute, Version 6, 2017.

[16] Royce, W. W., "Managing the Development of Large Software Systems: Concepts and Techniques," *WESCON Technical Papers,* Vol. 14, August 1970, pp. A/1-1–A/1-9.

[17] Forsberg, K., and H. Mooz, "The Relationship of Systems Engineering to the Project Cycle," *Proceedings of the NCOSE Conference,* Chattanooga, TN, October 1991.

[18] Davis, A. M., E. H. Bersoff, and E.R. Comer, "A Strategy for Comparing Alternative Software Development Life Cycle Models," *IEEE Transactions on Software Engineering,* Vol. 14, No. 10, October 1988, pp. 1453–1461.

[19] Dorfman, M., and R. H. Thayer, *Software Requirements Engineering,* Los Alamitos, CA: IEEE Computer Society Press, 2000, p. 9.

[20] Davis, A. M., E. H. Bersoff, and E. R. Comer, "A Strategy for Comparing Alternative Software Development Life Cycle Models," *IEEE Transactions on Software Engineering,* Vol. 14, No. 10, October 1988, pp. 1453–1461.

[21] Dorfman, M., and R. H. Thayer (eds.), "System and Software Requirements Engineering," *IEEE Computer Society Press Tutorial,* 1990, pp. 543–552.

[22] Boehm, B., "A Spiral Model of Software Development and Enhancement," *IEEE Computer,* Vol. 21, No. 5, May 1988, pp. 61–72.

[23] Martin, J. N, *Systems Engineering Guidebook: A Process for Developing Systems and Products,* Boca Raton: CRC Press, 1997.

[24] Manifesto for Agile Software Development, 2001, http://agilemanifesto.org.

[25] Project Management Institute, *Agile Practice Guide,* 2017.

[26] International Institute of Business Analysis, *Agile Extension to the BABOK Guide, Version 1,* Center for Systems Management.

[27] Project Management Institute, *Agile Practice Guide,* 2017.

List of Acronyms

ABL	Allocated baseline
ANSI	American National Standards Institute
AT&E	Acceptance test and evaluation
BA	Business analyst
BNR	Business needs and requirements
BRS	Business requirement specification
CCB	Configuration control board
CDR	Critical design review
CFD	Control flow diagram
CI	Configuration item
CM	Configuration management
CMP	Configuration management plan
CONOPS	Concept of operations
COTS	Commercial-off-the-shelf
CPM	Critical path method
CSC	Computer software component
CSCI	Computer software configuration item
CSU	Computer software unit

CWBS	Contract work breakdown structure
DDR	Detailed design review
DID	Data item description
DLOD	Defense lines of development
DoD	Department of Defense
DOTMLPF	Doctrine, organization, training, material, leadership, personnel, facilities
DSMC	Defense Systems Engineering Management College
DT&E	Developmental test and evaluation
EIA	Electronics Industries Association
EMC	Electromagnetic compatibility
EMI	Electromagnetic interference
ERD	Entity relationship diagram
FBL	Functional baseline
FCA	Functional configuration audit
FEA	Finite element analysis
FFBD	Functional flow block diagram
FIC	Fundamental inputs to capability
FMECA	Failure modes, effects, and criticality analysis
FQR	Formal qualification review
FRACAS	Failure reporting, analysis, and corrective action system
GOTS	Government-off-the-shelf
GPS	Global positioning system
HDBK	Handbook
HMI	Human-machine interface
HW	Hardware
HWCI	Hardware configuration item
ICD	Interface control document
ICWG	Interface control working group

IDEF	Integrated definition
IEC	International Electrotechnical Commission
IEEE	Institute of Electrical and Electronics Engineers
IIBA	International Institute of Business Analysis
IIS	Introduction into service
ILS	Integrated logistics support
INCOSE	International Council on Systems Engineering
IS	Interim standard
ISO	International Organization for Standardization
LC	Life-cycle concept
LCC	Life-cycle cost
LSC	Logistic support concept
MBD	Model-based design
MBSE	Model-based systems engineering
MIL-HDBK	Military handbook
MIL-STD	Military standard
MOTS	Military-off-the-shelf
MTP	Master test plan
NCOSE	National Council on Systems Engineering
NRM	(INCOSE) *Needs and Requirements Manual*
OOA	Object-oriented analysis
OpsCon	Operational concept
OT&E	Operational test and evaluation
OTS	Off-the-shelf
PBL	Product baseline
PCA	Physical configuration audit
PDR	Preliminary design review
PERT	Program evaluation and review technique
PLC	Preliminary life-cycle concept

PLM	Product life-cycle management
PM	Project management
PMI	Project management institute
PMP	Project management plan
PRICIE	Personnel, research and development, infrastructure, concepts and doctrine, information technology, equipment
PSTN	Public Switched Telephone Network
PWBS	Program/project work breakdown structure
QA	Quality assurance
QFD	Quality function deployment
RAAF	Royal Australian Air Force
RAN	Royal Australian Navy
RBS	Requirements breakdown structure
RF	Radio frequency
RFT	Request for tender
RMP	Risk management plan
RNZAF	Royal New Zealand Air Force
SBP	Strategic business plan
SBS	System breakdown structure
SDD	System design document
SDR	System design review
SE	Systems engineering
SEBoK	*Systems Engineering Body of Knowledge*
SEDS	Systems engineering detailed schedule
SEI	Software Engineering Institute
SEMP	Systems engineering management plan
SEMS	Systems engineering master schedule
SNR	Stakeholder needs and requirements
SOI	System of interest

SOP	Standard operating procedure
SoS	System of systems
SOW	Statement of work
SRD	Stakeholder requirement document
SRR	Systems requirements review
SSADM	Structured systems analysis and design
STD	Standard
StRS	Stakeholder requirements specification
SVR	System verification review
SW	Software
SWEBoK	Software Engineering Body of Knowledge
SyRS	System requirements specification
SysML	System modeling language
T&E	Test and evaluation
TCO	Total cost of ownership
TEMP	Test and evaluation master plan
TPM	Technical performance measure
TRAP	Technical review and audit plan
TRR	Test readiness review
UML	Unified modeling language
V&V	Verification and validation
WBS	Work breakdown structure
WSOI	Wider system of interest

About the Authors

Ian Faulconbridge holds a bachelor's, a master's, and a doctorate in engineering, an MBA in project management, and a master's degree in environmental management (in conservation). Dr. Faulconbridge is a Fellow of Engineers Australia (FIEAust), an Engineering Executive (EngExec), a Chartered Professional Engineer (CPEng) in four colleges, and a registered professional engineer. He has over 35 years of applied professional engineering practice and has performed senior systems engineering roles within the defense and aerospace sectors both in Australia and overseas. To augment and inform his professional practice, Ian has also spent periods as either a full-time or an adjunct engineering academic at the University of New South Wales, the Australian National University, the Queensland University of Technology, the University of Southern Queensland, and James Cook University. He has authored or coauthored seven books covering systems engineering, radar and electronic warfare, and avionics systems. Collectively, over 130,000 students have attended one of Ian's courses, and over 10,000 copies of Ian's texts have been sold.

Michael Ryan holds bachelor's, master's, and PhD degrees in electrical engineering. Dr. Ryan is a FIEAust, a CPEng in systems engineering, electrical, and ITEE colleges, a senior member of IEEE (SMIEEE), a Fellow of the International Council on Systems Engineering (FINCOSE), a Fellow of the Institute of Managers and Leaders (FIML), and a Fellow of the Royal Society of NSW (FRSN). He has over 45 years of experience in communications engineering, systems engineering, capability management, and management. From 1998 to 2020, he served as a professor at the University of New South Wales, where he was most recently the director of the Capability Systems Centre. His research and teaching interests are in communications and information systems, requirements engineering, systems engineering, and project management—he also regularly consults in those fields. He is a cochair of the Requirements Working Group (RWG) of INCOSE. He is the author or coauthor of 14 books, four book chapters, and over 450 refereed journal and conference papers.

Index

Artech House Technology Management and Professional Development Library

Seth Sullivan, Series Editor

Actionable Strategies Through Integrated Performance, Process, Project, and Risk Management, Stephen S. Bonham

Advanced Systems Thinking, Engineering, and Management, Derek K. Hitchins

Applied Systems Engineering, Second Edition, Ian Faulconbridge and Michael Ryan

Applying Total Quality Management to Systems Engineering, Joe Kasser

Building Successful Virtual Teams, Francine Gignac

Critical Chain Project Management, Third Edition, Lawrence P. Leach

Decision Making for Technology Executives: Using Multiple Perspectives to Improve Performance, Harold A. Linstone

Designing the Networked Enterprise, Igor Hawryszkiewycz

Electrical Product Compliance and Safety Engineering, Volume 2, Steli Loznen and Constantin Bolintineanu

The Engineer's Guide to Authentic Leadership, Ronald J. Bennett and Elaine R. Millam

Engineer's and Manager's Guide to Winning Proposals, Donald Helgeson

Engineering and Technology Management Tools and Applications, B. S. Dhillon

Enterprise Release Management: Agile Delivery of a Strategic Change Portfolio, Louis Taborda

The Entrepreneurial Engineer: Starting Your Own High-Tech Company, R. Wayne Fields

Evaluation of R&D Processes: Effectiveness Through Measurements, Lynn W. Ellis

From Engineer to Manager: Mastering the Transition, Second Edition, B. Michael Aucoin

Global High-Tech Marketing: An Introduction for Technical Managers and Engineers, Jules Kadish

How to Become an IT Architect, Cristian Bojinca

Integrated IT Project Management, Kenneth R. Bainey

Introduction to Information-Based High-Tech Services, Eric Viardot

Introduction to Innovation and Technology Transfer, Ian Cooke and Paul Mayes

ISO 9001:2000 Quality Management System Design, Jay Schlickman

IT Project Portfolio Management, Stephen S. Bonham

Managing Complex Technical Projects: A Systems Engineering Approach, Second Edition, Ian Faulconbridge and Michael Ryan

Managing Successful High-Tech Product Introduction, Brian P. Senese

Managing Virtual Teams: Practical Techniques for High-Technology Project Managers, Martha Haywood

Mastering Technical Sales: The Sales Engineer's Handbook, Fourth Edition, John Care

The New High-Tech Manager: Six Rules for Success in Changing Times, Kenneth Durham and Bruce Kennedy

The Parameter Space Investigation Method Toolkit, Roman Statnikov and Alexander Statnikov

Planning and Design for High-Tech Web-Based Training, David E. Stone and Constance L. Koskinen

A Practical Guide to Managing Information Security, Steve Purser

Practical Model-Based Systems Engineering, Jose L. Fernandez and Carlos Hernandez

Practical Reliability Data Analysis for Non-Reliability Engineers, Darcy Brooker with Mark Gerrand

The Project Management Communications Toolkit, Second Edition, Carl Pritchard

Project Managment Process Improvement, Robert K. Wysocki

Reengineering Yourself and Your Company: From Engineer to Manager to Leader, Howard Eisner

The Requirements Engineering Handbook, Ralph R. Young

Requirements Practice in Conceptual Design, Second Edition: A Practical Guide, Michael Ryan

Running the Successful Hi-Tech Project Office, Eduardo Miranda

Successful Marketing Strategy for High-Tech Firms, Second Edition, Eric Viardot

Successful Proposal Strategies for Small Businesses: Using Knowledge Management to Win Government, Private Sector, and International Contracts, Sixth Edition, Robert S. Frey

Successful Proposal Strategies On-the-Go!, Robert S. Frey

Systems Approach to Engineering Design, Peter H. Sydenham

Systems Engineering Principles and Practice, H. Robert Westerman

Systems Reliability and Failure Prevention, Herbert Hecht

Team Development for High-Tech Project Managers, James Williams

For further information on these and other Artech House titles, including previously considered out-of-print books now available through our In-Print-Forever® (IPF®) program, contact:

Artech House
685 Canton Street
Norwood, MA 02062
Phone: 781-769-9750
Fax: 781-769-6334
e-mail: artech@artechhouse.com

Artech House
16 Sussex Street
London SW1V 4RW UK
Phone: +44 (0)20 7596-8750
Fax: +44 (0)20 7630-0166
e-mail: artech-uk@artechhouse.com

Find us on the World Wide Web at: www.artechhouse.com

www.ingramcontent.com/pod-product-compliance
Lightning Source LLC
LaVergne TN
LVHW051045210926
848486LV00001B/31
9781685691318